Medical Device Product Development Process

ISBN: 9798431106125
Imprint: Independently published

© 2022

Forward

This book is intended to provide an introduction to the application of a lifecycle approach to product design and development for medical devices. Medical Device Product Lifecycle management provides a framework to develop, design maintain user requirements and ensure the safety and performance of medical devices. Application of a Medical Device Product Lifecycle Management benefits the business aspects of manufacturing, fosters alignment across design and development teams and incorporates the voice of the customer, taking into account their needs and safety that is inherent in the design of products.

The Medical Device Product Lifecycle (MDPL) relies upon several distinct quality management elements and processes in order to function effectively. These include; principles and establishment of a quality management system, regulatory processes, validation processes, engineering processes, change management and risk management processes. The MDPL process itself requires procedures and documentation to facilitate product realization.

For the reader to gain the maximum benefit from this book, the following points should be considered. To begin with-Regulations are mandatory. Medical devices range in their application (intended use and indications), technologies, principles of operation, complexity and value. However, regulation in addition to standards need to be consulted and applied within organizations. The manufacturer has a legal responsibility in this regard and the classification and commercial strategy may require regulations such as FDA 21 CFR 820, (United States), and Medical Device Regulations EU MDR (2017/745) or Regulation 2017/746 on In-Vitro Diagnostic Devices (IVDR) (in Europe).

This book provides a general framework for MDPL. -It does not intend to provide a *ready-made* or *drop-in* process for manufacturers. Device manufacturers themselves must establish and maintain appropriate design and development processes etc. For different classifications of medical devices, certain aspects of the D&D process may require greater focus, verification and validation. The knowledge of appropriately qualified and experienced Quality and Design Engineers, medical professionals and clinical experts needs to be drawn upon to input and review the project.

Contents

Contents

Forward..2
Medical Device Product Lifecycle Management Process...............................8
   21 CFR 820.30 Design Controls 8
   Introduction..8
   Process Overview 9
   Framework for Medical Device Product Lifecycle 10
   21 CFR 820.30 Design Controls 11
   General Concepts 13
      Integration of Design Controls 13
      Risk Management Activities 14
      Document Name 15
      Risk Management Activity 15
      Risk Management Plan 15
      Design Risk Analysis 15
      Use Related Risk Analysis 15
      A risk assessment tool that identifies potential use errors and mitigations 15
      Process Failure mode and Effects Analysis 15
      Risk Management Report 15
   Design Risk Analysis 18
      Design Reviews and Gate Reviews 19
      Design History File, DHF 19
-Stage One-..21
   Stage I – Ideation/Concept Evaluation 21
      Introduction..21
      Project Initiation 21
      Design History File (DHF) and Index (DHFi) 21
      Useful Definitions...22

  Core Team Formation 23
Stage II – Planning & Design Input 24
  Design Input Requirements for Product Development 24
  Introduction 24
  Design & Development Planning 24
  Stakeholder Needs 25
  Design Input Requirements for Product Development 26
21 CFR 820.30 (c) 26
  Essential Requirements 27
  Design Inputs Outputs Verification and Validation Matrix, (DIOV) ... 30
  Design Review 30
Design Change Control 31
  Design History File 32
Stage III – Development 33
  Introduction 33
  Development of Design Outputs 33
  Design Risk Management 34
  Design Verification Activities 34
Stage IV- Implementation 36
Design Validation 36
Stage V- Design Transfer 37
  Introduction 37
Process Validation 37
  Stages of Process Validation 38
  Fundamentals of Process Validation 38
  Process Operational Qualification (OQ-P) 39
  Process Performance Qualification 39
  Continued Process Verification 39
  Revalidation (or Maintaining a Validated State) 40
  Validation Strategies 40
  Principles of Worst Case Selection 40
  Requalification 40

  Types of Validation 41
Test Methods 43
  Introduction 43
  Factors to consider for Test Method Selection and Validation 43
  Definitions 44
Scenarios 46
  New Test Methods 46
  Changes to Existing Methods 46
  Method Transfer 46
  Ruggedness 46
  Key Concepts 46
  Protocols 49
  Test Method Accuracy 49
  MSA Studies 49
  General MSA Requirements: 50
  Variable MSA Studies : 51
  Attribute MSA Studies 52
  Non Destructive 52
  Destructive 52
  Measurement Capability Index 52
Risk Management for Product Lifecyle 53
  Risk Analysis 60
  Risk Estimation / Evaluation 63
  Risk Control 65
  Risk Acceptability 65
  Criteria for risk acceptability 66
  Risk management Review and Reporting 66
  Risk Management File 68
  Overall Residual Risk 69
  Post Production Review 70
  Role of Standards 71
Failure Modes And Effects Analysis 75

ISO 13485 Quality Management System for Medical Devices 80
Code of Federal Regulations, FDA, 21 CFR Part 820........................80
Factors in Device Design, Safety and Performance 81
ISO 13485 & Product Realization..82
Planning of Product Realization / Design and Development Planning 82
Failure Modes And Effects Analysis (FMEA)...............................82
Why FMEA? 84
Methodology for FMEA..84

Usability Engineering and Product Design and Development 88

Key Terms..90

Usability Engineering Process 94

Inputs into Risk Identification 95

Preliminary Evaluations 96

Critical Task Identification and Categorization 96

Identification of Known Use-Related Problems 97

Usability Engineering Process Overview 98

Use Related Risk Analysis 108

Formative Studies and Evaluation 112

Efficiency 115

Safety 116

Usability Engineering File 116

Use Scenario 117

Environmental Factors and Usability 117

Use Specification 117

User Interface 118
User Needs and Requirements 120
User Interface Evaluation Plan 121

Summative Testing 122
    Summative Evaluation Protocol 122
Post-Marketing Surveillance 124
European Regulations- Usability and MDR 126
    Design Controls and Usability Engineering 127
    Appendix- 4- Simple format of a Use Related Risk Analysis 131
    Information Supplied and Usability 134
Useful Definitions 137
APPENDIX - Risk Questionnaire 138
APPENDIX - Use Engineering and Product Development 148
APPENDIX MDR Annex I, General Safety and Performance Requirements 149
APPENDIX - Risk Management Plan..173

Medical Device Product Lifecycle Management Process

In this section:

> *Introduction*
> *Process Overview*
> *Framework for Medical Device Product Lifecycle*
> *21 CFR 820.30 Design Controls*
> > *Design Controls and ISO 13485 -Quality Management System for Medical Devices*
> *General Concepts*
> > *Integration of Design Controls*
> > *Risk Management Activities*
> > *Design Reviews and Gate Reviews*
> > *Design History File, DHF*
> > *Useful Definitions*
> > *Core Team Formation*

Introduction

Establishing a Medical Device Product Development Process is necessary to for medical device manufacturers to meet FDA regulations and regional regulations applicable to the products for sale. Beyond the legal stipulation, an additional purpose is to realize and deliver safe and effective medical devices that meet the intended use and user needs which provides medical solutions that enhance and preserve quality of life. To achieve safe and effective products, the guiding principles and techniques not only need to be applied in the design and development stages, but maintaining and monitoring the performance through the product lifecycle is required. It is easily to compartmentalize product performance and safety as a development and design-based activity, however, an effective medical device lifecycle process must ensure design safety and continuity until product retirement and discontinuation.

The introduction of new medical device products or changes to existing medical device products necessitate design and development activities in order to plan and deliver the appropriate verifications and validations to demonstrate safety and performance of products.

Device manufacturers in establishing and applying a Medical Device Product Lifecycle Process must continually ensure the process is fit-for-purpose and that is fulfils its legal and regulatory obligations (e.g. meeting the requirements of *21 CFR 820.30: Medical Devices - Quality System Regulation and ISO 13485: Medical Devices- Quality Management Systems etc.)*

Process Overview

A Medical Device Product Lifecycle process must be a comprehensive, end-to-end process that encompasses the entire life cycle of a product from project initiation through to product discontinuation and retirement of a marketed product. Making a comprehensive and detailed process intuitive and easy-to-apply should also play a role in designing and maintaining a MDPL process. Structuring the process into specific stages and the use of design review or gate reviews also provides for a format that has milestones and a stage by stage approach that makes it easier for engineers to work with.

The first stage of any MDPL process concerns the business case and what product is to be developed, selected and realized through the process itself. The other stages involve design, development, implementation, design transfer, product launch and lifecycle management.

Each stage is gated by a design review and before the project should move onto the next stage. While the MDPL is a continual process until product retirement, the initial product release and launch is achieved by successful design transfer and completion of D&DP requirements.

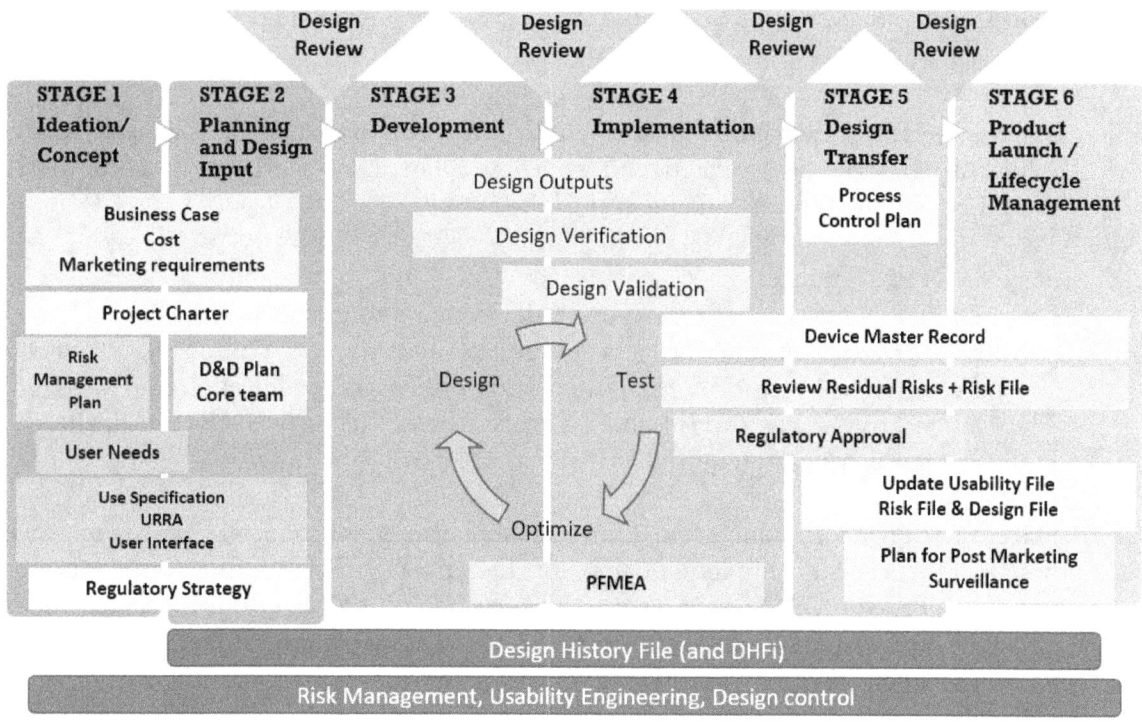

Framework for Medical Device Product Lifecycle

Stage I – Ideation / Concept Evaluation. For a new product or for significant redesign and relaunch, a business case must be established. This occurs at the beginning of design effort. Factors to consider include the technology, cost, manufacturing platform and technology, resources, market feasibility and market selection. These factors help to 'test' whether there is sufficient cause to design and develop a product. This stage therefore features key information that is required to resource, justify and outline a design project and is often summarized in a Project initiation document (PID).

Stage II – Planning & Design Input. Design control is typically initiated upon approval of the planning and design inputs and subsequent review report. To host the Planning and Design Inputs review, some core documents are required. This can include the *Design & Development Plan (D&DP)*, design input requirements (design inputs) based on stakeholder needs. The creation of a Risk management planning and a design risk analysis begins. A Design Review is required to approve this stage of the MDPL.

Stage III – Development. In development, design concepts are created to meet the design input requirements. Design verification testing is completed during this stage to confirm that the design input requirements have been met. Process development activities should also be progressed. A Design review is typically required to review and approve the progress against the D&DP.

Stage IV – Implementation. Design Validations are completed. Design and process validation strategies are created and approved. The Device Master Record (DMR) should be nearing completion and approval.

Stage -V-Design Transfer
Design transfer involves the translation of the design into production specifications and transferred to manufacturing. Process and design validations are completed. Production and process controls must be established and verified via qualification and validation. The Design History File (DHF) and a Device Master Record (DMR) are finalized. A Design review is typically required to review and approve the progress.

Stage VI – Product Launch and Lifecycle Management. Product release is approved based on successful implementation of the D&DP, verifications and validation approved, Process Validation approved all completion of Design transfer requirements. The Lifecycle requirement of products requires monitoring of Post production data and post-launch review.

21 CFR 820.30 Design Controls

Design controls are an interrelated set of practices and procedures that are incorporated into the design and development process. A key purpose of design control is to increase the likelihood that the design transferred to production will translate into a device that is appropriate for its intended use. The cost to correct design errors is lower when errors are detected early in the design and development process. Design control does not end with the transfer of a design to production, as it applies to the device or manufacturing process design, including those occurring long after a device has been launched.

Code of Federal Regulations, 21CFR 820.30, Subpart C - Design Controls
Sec. 820.30 Design controls.
    (a) General.
    (b) Design and development planning.
    (c) Design input.
    (d) Design output.
    (e) Design review.
    (f) Design verification.
    (g) Design validation.
    (h) Design transfer.
    (i) Design changes.
    (j) Design history file.

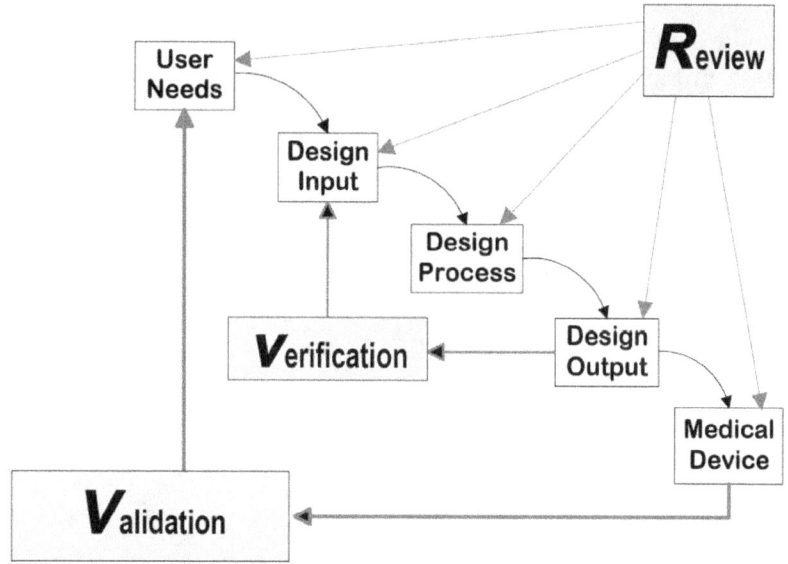

The role of design input and verification of design outputs is illustrated above. Each design input is converted into a new design output; each output is verified as conforming to its input; and it then becomes the design input for another step in the design process.

Medical devices exhibit different levels of complexity and risk, ranging from simple devices such as bandages, plasters and urine test strips to automated diagnostic devices, orthopedic implants, bone screws and so on. Similarly, manufacturing companies vary in size, structure, and while they must comply with regulatory requirements of design and development, their approach and management practices may vary.

Design controls are a requirement of quality systems such as 21 CFR Part 820 (medical devices), and for certain classes of devices and per ISO 13485 - Quality Management Systems.

Benefits of a Design Control:

1.0 The intended use of the device is documented and approved
2.0 It ensures inputs align with outputs
3.0 It creates a design "standard" and a "process" to allow benchmarking and consistency within an organization

Design Controls and ISO 13485 -Quality Management System for Medical Devices

Clause 7 of ISO 13485 specifies the requirements for design and development of devices as part of the product realization process. It should be noted that organizations can opt to exclude specific requirements of ISO 13485, in cases where product realization is not applicable. However, any such exclusion should be based on sound rationale with the technical case clearly documented. An example of this may be where design and development are not conducted by the manufacturer e.g. contract manufacturers.

Clause 7 (product realization) of ISO 13485 details requirements for design and development controls. Clause 7 includes the following subparts:

Clause 7.1 Planning of product realisation
Clause 7.2 Customer-related processes
Clause 7.3 Design and development
Clause 7.4 Purchasing
Clause 7.5 Production and service provision
Clause 7.6 Control of measuring devices

Section 7.3 (Design and development) comprises:

Clause 7.3.1 Design and Development Planning
Clause 7.3.2 Design and Development Inputs
Clause 7.3.3 Design and Development Outputs
Clause 7.3.4 Design and Development Review
Clause 7.3.5 Design and Development Verification
Clause 7.3.6 Design and Development Validation
Clause 7.3.7 Control of Design and Development Changes

Throughout this procedure those activities, procedures, and templates that specifically address a design control element will be indicated by the following symbol.

General Concepts

For most Medical Device Lifecycle processes some core concepts, are listed below that feature across the different stages of product design and development and provide the structure and tools to successfully complete a product development project.

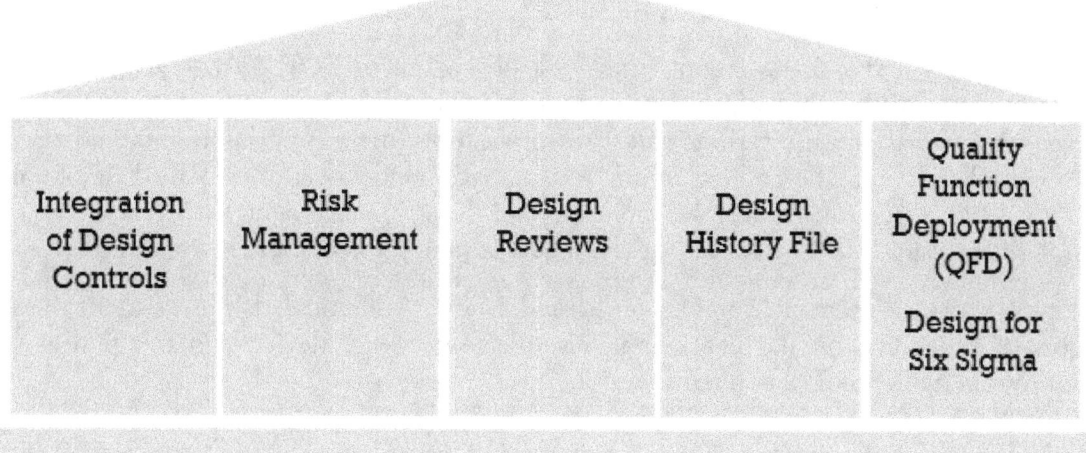

Integration of Design Controls

Design Controls must be integrated throughout lifecycle of products to ensure that product design continuity is maintained and processes are aligned with industry standards and comply with:
- 21 CFR 820.30 *Medical Device - Quality System Regulation* and
- ISO 13485 *Medical Devices – Quality Management Systems*

The Design & Development planning stage ensures that relevant design control activities are planned and implemented over the course of the project and at the appropriate stages.

Risk Management Activities

Risk management involves the systematic application of management policies, practices and procedures that identify, analyse, control and monitor risk.

It is important to recognise that risk management should begin at the outset of the design and development phase of a project. The first step is to identify the user needs and intended use and application of the device. At the design input phase and design selection phase, risk assessments should be in a mature state.

This allows the review of potential risks relating to the design of the product. Unacceptable risks can be dealt with by means of revisiting the design or introducing controls or mitigations in order to reduce the risks to acceptable levels.

Following on from the design and development phase, the design verification, validation and transfer phases, or the clinical readiness phase, risk management activities and acceptability of the residual risk become the focus and must be approved indicating acceptability. This is often referred to as communicated risk.

In order to apply a risk management strategy, a procedure or SOP on risk management is typically available within manufacturing companies. This should clearly describe the risk management process and the various risk assessment tools, their application and guidance on how to complete them. The content of any risk management procedure or SOP should align with ISO 14971 Medical Devices - Application of Risk Management to Medical Devices. Controlled templates for PFMEAs etc. also bring consistency and continuity to the process.

The key risk management deliverables and listed below. Additional risk assessment may be appropriate depending on the nature and classification of the device and the polices and procedures internally within a company.

Document Name	Risk Management Activity
Risk Management Plan	Details the risk management activities to be completed and details the methods, effectiveness criteria and data to be gathered and reviewed
Design Risk Analysis	A risk document that summaries the design related risks of a product and the control and mitigations required in order to lower the risk level to as low as possible and to acceptable levels.
Use Related Risk Analysis	A risk assessment tool that identifies potential use errors and mitigations
Process Failure mode and Effects Analysis	A risk assessment approach for processes where the various failure modes, their cause, effect and impact are assessed and scored. Usually estimating severity, occurrence and detection for each failure mode
Risk Management Report	A report where the output of the completed risk activities are summarised and conclusions drawn. Confirmation that the risk plan has been completed as planned with no deviations. Confirmation that residual risk is as low as possible and acceptable. Confirmation that the benefits outweigh the risks.

Risk Management Plan

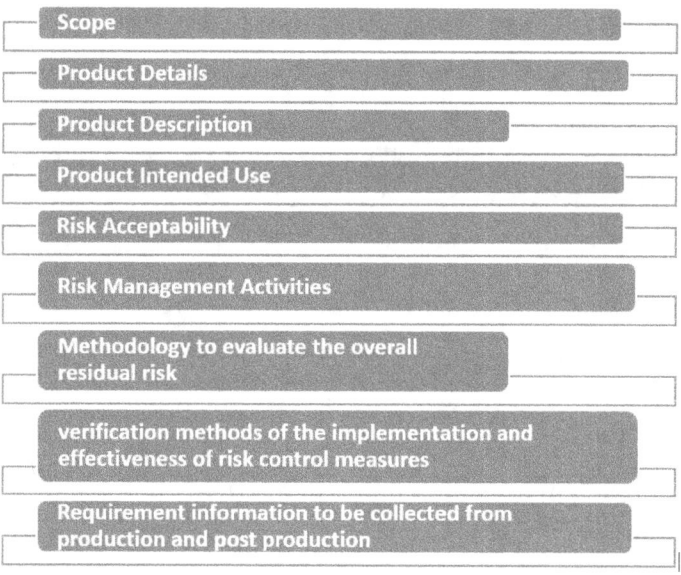

Scope:

The product within the scope of the project.

Product Details:

> Product name(s) (While it may seem obvious, some brands or products may have many similar but different products on the market. For example device may have varying levels of complexity and functionality pitched a different prices on the market, e.g. Product *Pro, Product *Gold, Product* extra.

Product SKU:

> The Risk management plan should set out the product or products included in the scope of the plan. Sometimes one plan will cover a family of products or subgroup or products. This practice is common and acceptable. Defining the scope of the risk management plan is a requirement of ISO 14971.

Product description:

> A product description should be informative and consistent between marketing materials, patient literature and design documentation.

Intended use:

> The intended use of the product must be included in the Risk management plan.

Roles and Responsibilities:

Roles and responsibilities and required to be included per ISO 14971. Defining such roles and responsibilities ensures the right expertise and invested in risk management and helps to ensure successful application of the methods and verifications that ensure an effective risk management process. The plan must identify all of the Risk Management activities planned during the lifecycle of the product. These include:

<u>Risk Management Plan inputs</u>

- Management Review
- Design Management
- Risk Management
- Change Management
- Complaint Handling
- CAPA Management
- Clinical Evaluation Reports
- Adverse reportable events

The risk management plan details by what means (how) and when (new products, product design changes, periodic review) the risk management activities will be reviewed for a medical device or defined family. Key sections within a compliant risk management plan include the method of review, the responsible individuals/ functions, how the output of the review is managed. Results are then reflected in the risk management report.

Criteria for risk acceptability

The ISO 14971 Risk management process requires a manufacturer's to have a policy that requires the establishment of what is deemed acceptable risk- also known as risk acceptability. To ensure this policy is established with the right parameters and for it to remain impartial, the criteria for risk acceptance should be created before commencing the risk assessment.

Risk Acceptability

The manufacturer must have a policy for determining acceptable risk, including criteria for accepting risks when the probability of occurrence of harm cannot be estimated. These requirements should be included in the plan or a reference to the applicable document.

Method to evaluate overall residual risk and criteria for acceptability

The overall residual risk and the criteria for its acceptability are based on the manufacturer's policy for establishing criteria for risk acceptability. Per ISO 14971, the method and the criteria must be stated in the risk management plan for the particular medical device.

Verifications methods and activities

The risk management plan should specify how the verification activities are executed, or alternatively it should reference another document that provides the details. The plan should specify the methods that are used to verify that any risk Control measures are implemented and to what degree they are effective. In addition, the overall residual risk must have a method of evaluation and criteria for the acceptability of the overall residual risk.

Post production and Post Marketing Requirements

Post-production information becomes an input into Risk Management activities for the product and also features in the risk management plan. The type of post marketing surveille (sources of information, analysis of information) should be appropriate for the product covered in the plan.

Design Risk Analysis

Design risk analyses' should be utilized from the planning stage of a project and be applied throughout the design and development process to ensure proper mitigation of identified risks.

The Design Risk Management plan should be prepared and approved according to the requirements ISO 14971. ISO TR/ 24971 also contains useful risk tools in the assessment of design risk.

Risk Management Policy

The purpose of risk management is to ensure that design risks are identified early enough in the process such that these issues may be assessed and mitigated as early as possible.

An overarching risk management policy must be established in advance of risk management activities that sets out the approach to risk and consider how risk acceptability is applied.

Usability and Risk Management

Usability testing or usability engineering studies can be performed during the development of a new product. It acts as a verification that a device is designed appropriately and can identify scenarios or conditions that users could present a use error or usability risk to the patient or user.

IEC 62366-1 Medical Devices-Application of usability engineering to medical devices is referenced both in ISO 14971 and ISO/TR 24971. While Risk Management and Usability Engineering are separate processes, they both supplement and overlap in their intent.

As defined above, Use Error is defined as a *"user action or lack of user action while using the medical device that leads to a different result than that intended by the manufacturer or expected by the user"* Technical report, ISO 24971:2020. This covers the following errors:

-the inability of the user to complete a task.

-Use errors resulting from a mismatch between the characteristics of the user, user interface, task, or use environment. Users may be aware or unaware that a use error has occurred.

Exception (to a use error):

- An unexpected physiological response of the patient is not by itself considered use error.

- A malfunction of a medical device

Identification of hazards from use errors

The usability testing or studies can highlight if issues occur when the device is used by the patient- for example, do people use the medical device in a way that it is not intended to be used or not in accordance with the instructions for use.

Hazards from reasonably foreseeable misuse

Some hazards and hazardous situations may be a result of reasonably foreseeable misuse. Engineering usability studies can also help identify and confirm reasonably foreseeable misuse scenarios.

Design Reviews and Gate Reviews

Design reviews and gate reviews track the progression of design projects over the course of its lifetime, but also acts as a checkpoint and readiness review for a project to proceed from one stage to the next one.

Key goals

1. provide feedback to designers on existing or emerging problems
2. assess project progress against the project D&DP.
3. provide confirmation that the project is ready to move on to the next phase of development

Design reviews must be completed to comply with the requirements 21 CFR 820.30(e) and ensure that the design is ready to move forward in the design control process.

There are three types of design reviews:
- Design Review of Inputs,
- Design Review of Outputs and Verification, and
- Manufacturing Readiness Design Review.

The term "phase approach" is often used when describing the design control process and how design reviews are built-in. It simply means that a sequence of tasks needs to be completed, reviewed and approved during the development cycle of a product or medical device. Tasks are grouped into phases or stages.

Design History File, DHF

The design history file contain the records that are necessary to demonstrate that the design was developed in accordance with the approved design plan with appropriate verification and validations. In real terms, the design history file is a compilation or collection of documents that describes the entire design history of a finished medical device product. The content includes the following:

- Approved Design Planning Documents
- Design Input Output Verification and Validation Documents
- Design verification and validation reports
- Design reviews reports
- Clinical Reports
- Risk Documentation

Design History File Index:
The collection of documents is the DHF, an index of listing of these documented and there storage location is often summarized in a Design History File Index (DHFi).

Document Title	Document Number	Location	Author	Revision	Effective Date

-Stage One-

Stage I – Ideation/Concept Evaluation

In this section:

Stage 1 Introduction
Project Initiation
Design History File
Useful Definitions to get started
Core team formation

Introduction

Stage I deals with the generation of ideas and concepts that provide the basis and starting point for a new product. Concepts should be reviewed by cross functional teams and feasibility demonstrated early-on. A concept document can be created to outline the product and applicable details. Commercial and marketing requirements should be understood along with the target price or average sale cost of the product. Upon the agreement of an organization to pursue a new medical device product, project initiation or a project charter should be created and approved by the key business units and functional stakeholders.

Project Initiation

Establishing a procedure around Ideation/ Concept Evaluation ensures that projects commence on a firm footing with consistency and core documents created. Once the business need and opportunity and the product is technically feasible, the project initiation process should capture key information, user requirements, marketing and business requirements.

Manufacturers, according to company requirements may establish strategic decision packages or other forms of documents to record approval and the details referred to above.

Design History File (DHF) and Index (DHFi)

Establishing the DHF is necessary in providing a common location for project documents that are necessary in demonstrating the design and development of the product. The DHF and DHFi should be established early-on to set the expectations for the team.

Useful Definitions

Design (Stage) Review: a process of evaluating the design requirements against the ability of it to deliver the intended device.

Design History File (DHF): an approved list of records that describe the design history of a medical device.

Design Input: the physical and performance requirements of a device that are the basis for the device design.

Design Output: the results of a design effort at each design phase and at the end of the total design effort. The finished design output is the basis for the device master record. The total finished design output consists of the device, its packaging and labelling, and the device master record.

Design Verification: confirmation by examination and provision of objective evidence that specified requirements have been fulfilled.

Design Validation: establishing by objective evidence that device or product specifications conform to user needs and intended use(s) defined in design documentation.

Device Master Record (DMR): a compilation of records containing the procedures and specification for a device. The contents of a DMR can contain local procedures such as SOPs and work instructions along with global or divisional specifications used to detail manufacturing processes, intermediate product or final product.

Design Phase Review: a documented, comprehensive, systematic examination of a design to evaluate the adequacy of the design requirements, the capability of the design to meet those requirements and to identify problems.

Specification: specification means any requirement to which a product, process, service, or other activity must conform.

Core Team Formation

Core Team should be formed when concepts move towards project initiation. Core teams are an essential element of project planning and the success of a design and development project. Some recommended minimum functions include:

- R&D functions
- Commercial
- Project Management
- Regulatory Affairs
- Medical Experts
- Clinical Affairs
- Engineering
- Supply Chain/Logistics
- Operations
- Quality
- Manufacturing/ Operations

The Core Team Roster shall be updated and approved whenever personnel and/or functions change.

Stage II – Planning & Design Input

In this section:

> *Introduction (Stage II)*
> *Design and Development Planning*
> *Stakeholder needs*
> *Design Input Requirements for Product Development*
> *Essential Requirements*
> *Design Inputs Outputs Verification and Validation Matrix*
> *Design Review*
> *Design Change Control*

Introduction

The key deliverable for Stage II is the creation of a *Design & Development Plan*. In addition to tasks, studies, documentation deliverables and timelines, the stakeholder needs are translated into design inputs.

Feasibility activities, such as technical feasibility and marketing studies, are not subject to design control requirements. However, the results of these activities will be reviewed if they form the basis for stakeholder needs or design input requirements.

In most scenarios, formal design control commences with the initiation of Design & Development planning review. Design Change Control should also be enforced after the approval of the *Design & Development Plan*.

Design & Development Planning

CFR 820.30(b)

> **Design input.** Each manufacturer shall establish and maintain procedures to ensure that the design requirements relating to a device are appropriate and address the intended use of the device, including the needs of the user and patient. The procedures shall include a mechanism for addressing incomplete, ambiguous, or conflicting requirements. The design input requirements shall be documented and shall be reviewed and approved by a designated individual(s). The approval, including the date and signature of the individual(s) approving the requirements, shall be documented.

The D&DP should cover the following at a minimum:

- Purpose and scope of the DDP
- Description of the project
- Resources Overview
- Key activities, verification, validations and timing
    - Description of activities with responsible person, delivery time
    - Activities should address risks identified in risk documents
    - Activities may be necessary for each Design input
- Timeline of Design reviews planned
- The D&DP is approved by core team members

Stakeholder Needs

Prior to establishing design inputs, stakeholder needs assist in shaping the product and assuring that the intended use, target market and other key requirements and identified. Stakeholders include end users such as patients or doctors or caregivers etc. The identification of stakeholder needs drives design input development.

Design Input Requirements for Product Development

21 CFR 820.30 (c)

> **Design input.** Each manufacturer shall establish and maintain procedures to ensure that the design requirements relating to a device are appropriate and address the intended use of the device, including the needs of the user and patient. The procedures shall include a mechanism for addressing incomplete, ambiguous, or conflicting requirements. The design input requirements shall be documented and shall be reviewed and approved by a designated individual(s). The approval, including the date and signature of the individual(s) approving the requirements, shall be documented.

The aim of the Design Input stage is to (1) define, identify and document the user needs, the intended use and other design criteria, materials and process requirements of the medical device. These are broadly known as stakeholder needs and (2) translate these stakeholder needs into specific (SMART) design input requirements. Design input requirements cover many technical areas, performance, safety, testing, regulatory and commercial requirements such as:

- Intended use
- Indications for use
- Marketing claims
- Performance and safety requirements,
- Physical characteristics,
- Human factors
- Biocompatibility & toxicity requirements
- Compatibility requirements (accessories)
- Packaging and labelling regulatory requirements of intended markets,
- Sterility requirements,

The typical documents required when establishing design inputs include:

- The creation of a formal design description detailing the intended use, user requirements and design inputs. (Note: the design description must align with the design input requirements.). This document is often referred to as a Design Inputs outputs verification and validation Document (DIOV).
- A design and development plan which provides an estimation of timelines, resources required, responsibilities, project risks and scope of the project, and accounts for actions and activities that shall fulfil the design inputs via testing and data
- Initial risk assessment which contains the user, design and component risks to be mitigated.
- Business case addressing the market size and market opportunity.

<u>Essential Requirements</u>

Standards within industry play an important multi-faceted role. In simple terms, and by their very nature, they encourage standardization across the globe in terms of the industry and individual products that differ in complexity. Standards are available for many types of medical devices. IVD devices, implants, simple medical devices and software powered complex equipment. The essential principles of device safety & performance of were first and foremost created by the Global Harmonization Task Force, (GHTF) and now archived by the International Medical Device Regulators Forum, IMDRF.

Understanding the regulatory viewpoint of standards is important in their use and the impact on product and process. Some standards may be voluntary but may inform manufacturers on how to comply with legislation. Some standards are mandatory and are required by regulatory legislation. In these cases, compliance to the status often meets the intent of the legislation.

ISO 16142-1, Medical devices – Recognized essential principles of safety and performance of medical devices -*Part 1: General essential principles and additional specific essential principles for all non-IVD medical devices and guidance on the selection of standards.*

Six (6) essential principles provide important fundamental criteria for design inputs that can result in an impact to risk management and the risk profile of a medical device.

Essential Principles of Safety and Performance
1.The medical device should be: -designed and manufactured in such a way that, when used under the conditions and for the purposes intended and, where applicable, by virtue of the technical knowledge, experience, education or training and the medical and physical conditions of intended users, they -**perform as intended** by the manufacturer and not compromise the clinical condition

or the safety of patients or users -risks which may be associated with their use constitute **acceptable risks when weighed against benefits** to the patient and are compatible with a high level of protection of health and safety.
a) reducing, as far as reasonably practicable and appropriate, the **risk of use error** due to the design of the medical device user **interface** and the **environment** in which the medical device is intended to be used
b) consideration of the technical knowledge, experience, education and training and, where applicable, the medical and physical conditions of intended users

Essential Principles of Safety and Performance
2. The solutions adopted by the manufacturer for the design and manufacture of the medical device should conform to safety principles, taking into account the **generally acknowledged state of the art**. When risk reduction is required, the manufacturer should control the risks so that **the residual risk** associated with each hazard is judged acceptable. The manufacturer should apply the following principles in the priority order listed below:
a) identify known or **foreseeable hazards** and estimate the associated risks arising from the intended use and foreseeable misuse
b) eliminate risks as far as reasonably practicable through inherently safe design and manufacture
c) reduce as far as reasonably practicable the remaining risks by taking adequate protection measures, including alarms or information for safety
d) inform users of any **residual risk**

Essential Principles of Safety and Performance
3. The medical device should achieve: -the **performance intended** by the manufacturer -**designed, manufactured and packaged** in such a way that during normal

conditions of use, they are suitable for their intended purpose

Essential Principles of Safety and Performance

4. The characteristics and performances referred to in essential principles 1, 2, and 3 (ABOVE) should not be adversely affected to such a degree that the health or safety of the patient/ user when:

*-the medical device is subjected to the **stresses which can occur during normal conditions of use** and has been properly maintained in accordance with the manufacturer's instructions.*

Essential Principles of Safety and Performance

5. The medical device should be:

designed, manufactured and packaged in such a way that their characteristics and performances during their intended use

will not be adversely affected by **transport** and **storage** conditions

(for example, fluctuations of temperature and humidity) taking account of the instructions and information provided by the manufacturer

Essential Principles of Safety and Performance

6. Any undesirable side-effect shall constitute an **acceptable risk** when weighed against the **performances intended**. All known and **foreseeable risks**, and any undesirable effects, should be minimized and be acceptable when **weighed against the benefits** of the intended performance of the medical device during **normal conditions of use.**

Design Inputs Outputs Verification and Validation Matrix, (DIOV)

DIOV combines the design inputs and design outputs in a common document or matrix. As described previously, the design input requirements are agreed and approved during the planning stage of a project. Each input must have an associated output which is the basis of ensuring requirements are specified in product specifications. During development and implementation design verifications and design validation reports are then included for each design requirement. Thus creating a powerful and important document that records critical design requirements and the evidence demonstrating effectiveness.

Design Review

21 CFR 820.30(e)

> **Design review**. Each manufacturer shall establish and maintain procedures to ensure that formal documented reviews of the design results are planned and conducted at appropriate stages of the device's design development. The procedures shall ensure that participants at each design review include representatives of all functions concerned with the design stage being reviewed and an individual(s) who does not have direct responsibility for the design stage being reviewed, as well as any specialists needed. The results of a design review, including identification of the design, the date, and the individual(s) performing the review, shall be documented in the design history file (the DHF)

Design Reviews must be a documented and comprehensive analysis of a design in order to evaluate the adequacy of the design requirements, to evaluate the capability of the design to meet those requirements, and to identify problems.

The full core team should be involved in reviews that ensure all functions are part of the project and the design review. An independent reviewer with appropriate technical expertise for the review should also be part of the design review. Design reviews satisfy both technical and regulatory requirements.

The purpose and scope of design reviews should cover the following:

- provide a documented review and examination of the design or candidate product
- review and update the design risks and mitigations identifying additional or emerging risks
- confirm the design is ready to move to the next stage in the MDPL
- Detect design deficiencies
- Status of documents, studies and verifications
- Revised planning if required

One independent must be present at all design reviews; the design review documentation must include justification in respect of the reviewer's independence and expertise. The Independent Reviewer shall not have direct responsibility for the design under review.

Design Change Control

During the development process while under design control, changes may need to be made. This may be due to information gained from engineering studies or verification studies, or modifications to allow for better manufacturability. Changes must be assessed against the D&D Plan, design inputs, outputs and regulatory strategies to name the key areas of potential impact. Therefore, a cross functional team should approve design change controls during development.

A separate Change Management process normally applies to products manufactured for commercial sale (products post design transfer) and should follow procedures that assess changes and the impact on the product.

21 CFR 820.30(i)

> Design changes. Each manufacturer shall establish and maintain procedures for the identification, documentation, validation or where appropriate verification, review, and approval of design changes before their implementation

Project teams and in particular quality representatives must manage design changes throughout the product life cycle. A method of *Design Change Control* must be used to approve the design change or changes and specify the actions or resulting updates to project documents as a result of the change.

The Design Change Control shall be used for initial approval and any subsequent changes to the following documents, at a minimum:

- Design & Development Plan (DDP)
- Stakeholder Needs Document or DIOV
- Design History File Index (DHFI)
- Risk Management File
- Usability File
- Design Verification impact
- Product Development Summary Report
- Design Validation Strategy
- Process Validation Strategy
- Quality Control Plan

Design History File

> **Design history file.** Each manufacturer shall establish and maintain a DHF for each type of device. The DHF shall contain or reference the records necessary to demonstrate that the design was developed in accordance with the approved design plan and the requirements of this part.

A Design History File Index (DHFI) must be established, which references all design documents and the appropriate document storage locations. A good rule of thumb is any document (verification/validation) contained in the DIOV should be recorded in the DHFi.

-Stage Three-

Stage III – Development

In this section:

>*Introduction*
>*Development of Design Outputs*
>*Design Risk Management*
>*Design Verification Activities*

Introduction

For the Development stage, further specification and iterations of the selected design concept(are conducted. As the design progresses, design outputs that fulfill design inputs continue to be developed. Process development must also commence and consider the process inputs and outputs required to manufacture the product.

The methodology of Concept, Design, Optimize is an effective way to improve the model and respond to any technical issues that are observed.

The specifics of design outputs can be subject to change during the development and the iterative nature of engineering, design and development. The design and development team must be cognizant of the design outputs and appropriate nominal values and tolerances for the product design and also the process parameters.

As new tasks are identified, the Design & Development Plan should be updated to remain accurate and should follow approval requirements in accordance with procedures.

Development of Design Outputs

Product specifications (PS) and the device master record (DMR) are generated based on design outputs.

Examples of Design Outputs are:

- Product specifications
- Process and Manufacturing specifications
- Component and material specifications
- Block diagrams
- Software High-level code
- Engineering drawings
- Software code
- Work instructions
- Quality specifications and procedures
- labeling specifications
- Test method specifications

Design outputs must cover the acceptance criteria and shall identify those design characteristics that are essential for the safe performance of the product.

Design Risk Management

21 CFR 820.30(g)

Design risk management activities must be maintained throughout the project. Additional design risk analysis activities may be conducted as required (e.g., Design Failure Mode & Effects Analyses, Fault Tree Analyses etc.). The goal of design risk management is to drive design choices that will mitigate risks to an acceptable level. If new design risks are identified that require mitigation, the design input requirements documentation shall be reviewed to determine whether a new design input requirement(s) is needed.

Design Verification Activities

21 CFR 820.30(f)

> **Design verification.** Each manufacturer shall establish and maintain procedures for verifying the device design. Design verification shall confirm that the design output meets the design input requirements. The results of the design verification, including identification of the design, method(s), the date, and the individual(s) performing the verification, shall be documented in the DHF.

Verification consists of the activities that confirm that design outputs meet design inputs. This is accomplished by examination and provision of objective evidence that specified requirements have been fulfilled. Conformance of inputs to outputs is the key focus. Design verification shall be performed on approved design outputs.

Examples of design verification tasks include:

1. Physical (fit, form) and functional tests
2. Chemical characterization
3. Microbial testing
4. Stability Studies
5. Package integrity tests
6. Biocompatibility testing
7. Failure modes and effects analysis
8. Bio-burden testing of products
9. Analytical testing
10. Environmental testing
11. Functional Testing
12. Usability Testing
13. Software testing
14. Toxicity testing
15. Sterility testing
16. Design failure modes and effects analysis (DFMEA)
17. Tolerance Stacking

-Stage Four-

Stage IV- Implementation

Design Validation

21 CFR 820.30(g)

> **Design validation.** Each manufacturer shall establish and maintain procedures for validating the device design. Design validation shall be performed under defined operating conditions on initial production units, lots, or batches, or their equivalents. Design validation shall ensure that devices conform to defined user needs and intended uses and shall include testing of production units under actual or simulated use conditions. Design validation shall include software validation and risk analysis, where appropriate. The results of the design validation, including identification of the design, method(s), the date, and the individual(s) performing the validation, shall be documented in the DHF

Project teams needs to develop the strategy for validating a design that meets the intended use but also demonstrates the product is safe, efficacious and meets the performance requirements. Design validations should be completed with the finished product to test the device in real world settings. Design validations therefore include clinical studies to demonstrate product validation.

A Design Validation plan can be developed to specify the activities required. For example, multiple design validations may be required to support specific regulatory requirements for regions or countries. Design validation, in similar fashion to verifications provide evidence that design inputs are addressed and the design outputs are suitably specified.

Usability testing such as summative evaluations can also be deemed design validations. Usability ensures that products are designed to minimize user errors.

-Stage Five-

Stage V- Design Transfer

In this section:

> *Introduction*
> *Process Validation*

§ 21 CFR 820.30(h)

> **Design transfer.** Each manufacturer shall establish and maintain procedures to ensure that the device design is correctly translated into production specifications.

Introduction

As the design output is finalised, the design is transferred into production specifications (drawings, manufacturing, test, and inspection procedures). Production specifications must ensure that manufactured devices are consistently and reliably produced within product and process capabilities, meeting all quality requirements.

Process Validation

Process validation is a statutory and regulatory requirement for the manufacture of medical devices. Per FDA 21 Code of Federal Regulations process validation is a regulatory requirement of Good Manufacturing Practices (GMP) for both pharmaceuticals (21 CFR 211) and medical devices (21 CFR 820). In addition to the regulatory drivers, process validation is a requirement in order to obtain certification to international standards issued by many notified bodies. (E.g. ISO 13485 Medical Devices – Quality Management Systems, ASTM E2500- Standard Guide for Specification, Design, and Verification of Pharmaceutical and Biopharmaceutical Manufacturing

Process-Operational Qualification (OQ-P)
The ability of a process to produce product in accordance with pre-determined specifications under worst case conditions. PQ is only required if no worst case conditions are evident.

Process-Performance Qualification (PQ)
The ability of a process to consistently produce product in accordance with predetermined specifications under anticipated conditions (normal/routine conditions). Before considering process validation in further detail, it is important to look at the prerequisites and other supporting activities required. These are examined in the sections below.

Stages of Process Validation

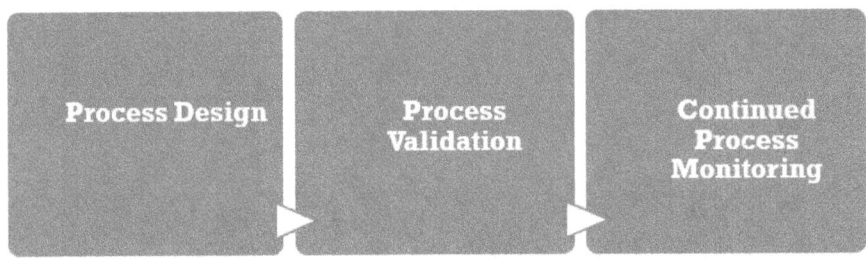

The three stages of process validation include: (1)Process Design, (2) Process Validation (3)Continued Process Monitoring. The commercial manufacturing process must be established during the process design phase. Key questions that need to be answered include: a)Definition of process inputs, b)Effects of inputs, c) Process outputs – CQAs (critical quality attributes) , d) Establishing process windows and risk assessment via DFMEA /PFMEA (design/process)

The process qualification stage looks at the validation of process design to confirm process is operating as intended and is capable of consistently producing product to meet quality requirements. Finally, stage 3, Continued Process Verification provides ongoing assurance through review of the process and quality metrics that the process is in control.

Fundamentals of Process Validation
Process validation is confirming that a process is capable of consistently manufacturing product under anticipated conditions. Remember, validation should be representative of the commercial process, so any issues in process validation will be repeated in commercial manufacturing.

The most important point when it comes to validation is that validation is neither exploratory nor investigative. Equally, it is not an engineering study. If you are ready to validate a system or process, all of the groundwork must be completed. This means critical parameters must be defined and documented, with technical rationale on why such parameters are critical etc. This body of work is typically done during a process development study or protocol.

Consistency, a core principle of process validation, is typically demonstrated by producing three batches/runs for a Process Performance Qualification (PPQ). These batches should be representative of normal production i.e. the size of the batch should be typical of commercial volumes. The PQ study should be executed at nominal conditions, (often termed "anticipated conditions") essentially referring to a controlled environment. Controlled material and controlled parameters (CPPs) are required. Nominal settings should be selected for PQ.

Process Operational Qualification (OQ-P)

During the Operational Qualification-Process (OQ-P) study, worst-case process conditions are normally employed. This may be worst case temperatures, speeds, feeds etc. The OQ-P should challenge the manufacture/processing of product at the limits of the processing window. If no worst-case conditions exist, then an OQ may not be required and only a performance qualification is required.

A family or matrix approach is often used where similar products are to be validated. A particular product size of product configuration may be selected to represent the worst-case product. Therefore, by qualifying the worst case, all other products within that family of products would be considered validated. However this approach must be clearly documented and technical rationale provided in advance of any qualification activities. This can be addressed in a validation plan or within a protocol.

Process Performance Qualification

The purpose of the PPQ is to demonstrate the capability of the process to consistently manufacture product to pre-determined specifications under normal operating conditions and defined parameters.

- Lots must meet the acceptance criteria set out in the protocol
- The lot size should be reflective of the intended lot size and also take into account normal variation
- If a family approach or matrix approach is used, the product selection must be clearly justified and documented
- Execute under anticipated conditions; essentially this refers to a controlled environment. Controlled material, controlled parameters (CPPs)
- Nominal settings should be selected for PPQ

Continued Process Verification

Once the initial validation is completed it is important that the system or process remains within the validated state, meaning that the system remains in a state of controlling process systems that capture information and data about the performance of the process. The use of statistical trending techniques should be considered. Data analysis of process and product should also include trending of raw materials, components and finished product. The purpose of process monitoring is to ensure critical parameters remain within control limits. It also helps to identify increasing variability or instability within the process which can then be investigated. All processes must have an upper and lower limit. If a process parameter only has a one-sided limit, then provide rationale in the OQ protocol to justify why a one-sided parameter window is acceptable. This requirement is not applicable to parameters that are set points.

Revalidation (or Maintaining a Validated State)

Revalidation is sometimes required if the original validation is no longer valid or representative of the process. Some instances where revalidation must be considered include changes to the process that can affect the product quality or efficacy, a removal, or the addition of a processing step or transfer of the equipment to a different location. In many companies an impact assessment is conducted if there is a proposal to modify a manufacturing process. Some changes may not require any validation while others may require a verification run.

When changes are proposed to the validated state of a process, the proposed changes must be fully understood in terms of the impact to product quality and the validated state. A risk assessment should be conducted to determine risks and appropriate mitigations.

Validation Strategies

A Family Approach (a.k.a. Bracketing, Matrix Approach) to validation is often used where a variety of similar products are manufactured using the same equipment. For process validation, a product that is representative of the family or group of products may be selected. Alternatively, a 'worst case' product may be selected as it presents the greatest challenge to manufacture to product specifications.

Principles of Worst Case Selection

Worst Case is a particular condition, set of conditions, and/or set of process parameters, generally made up of processing limits. Worst case conditions present the greatest chance of process issues or the greatest chance of failures due to product quality. Worst case conditions are used at the OQ-P stage to provide the greatest level of challenge, however, this is outside of normal operating conditions.

Requalification

During the lifetime of a process or piece of equipment, the need to re-qualify may arise. Such need should be assessed according to a validation procedure. Generally, the same tools used in the original validation can be re-applied to identify the need to re-qualify and indicate what requirements must be included.

The first step must be a review of the existing qualification, as changes may not impact the validated state, or may only require a limited requalification. For example, moving a piece of equipment may only require requalification of the utilities such as compressed air or process water if the operation of the equipment is not impacted by the movement and re-siting. Some examples where re-qualification may be required include:

- Changes to the process settings which may impact the product quality

- Change in raw material grade or vendor
- Changes to the design of the product
- Transferring a process from one plant to another plant
- Changes to manufacturing aids (e.g. cleaning agents, jigs and fixtures)

Types of Validation

The FDA provides clear definitions on the four types of validation which are explained below.

Prospective Validation

Establishing documented evidence **in advance** of process implementation that a process or system operates as intended. This is the preferred approach and is most common when new products must be validated before commercial manufacturing.

Concurrent Validation

Establishing documented evidence that a processes operates as intended, based on information generated during process implementation. Concurrent means that the outputs and performance of the system are monitored at the time of manufacturing which can include commercial lots.

Retrospective Validation

Retrospective validation is used for facilities or processes that have not completed formal validation. Historical data or a retrospective review can provide the evidence that the process or facility is operated as intended. This type of validation is uncommon.

Revalidation

Revalidation involves the re-execution of validation activities in order to maintain a validated state. This can be a result of substantial changes to product attributes, specification or changes to the manufacturing process itself. Other reasons a partial or full revalidation may be required involve instances where product quality issues have increased.

Design Transfer Documentation

Design Transfer principally depends upon the implementation of a Device Master Record (DMR). The DMR contains the specifications and instructions on how to manufacture, inspect, test and release the product. It can make reference to independent documents that contain the details and technical information. For example, a DMR may reference product specifications, raw material specifications and manufacturing specifications. Both the DMR and specifications should be controlled per GDP requirements and be approved by the appropriate team.

For the successful Design transfer of a product from the design and development stages to commercial manufacturing, a summary design transfer report may be utilized. In addition, the requirements of design reviews also apply to Design transfer stage. A design review must be held in order to approve the design transfer and ensure all aspects of the project have been successfully completed.

Test Methods

In this section:
- *Introduction*
- *Factors to consider for Test Method Selection and Validation*
- *Definitions*
- *Scenarios*
- *Ruggedness*
- *Key Concepts*
- *MSA Studies*
- *Measurement Capability Index*

Introduction

Test method validation involves the formal documentation of a test method used to capture and analyse data or information. Test methods are used to assess product outputs such as dimensions, material strength, chemistry and product functionality. Test Method validation and suitability is a critical element of the MDLP. Inspection and testing form the basis of informed decision-making and data-driven decisions. Therefore, their reliability and accuracy must be assured in order to support the right decisions such as specification setting and tolerance setting.

Factors to consider for Test Method Selection and Validation

(1) in-house requirements, these are internal requirements included in company policies and procedures that define the test method validation process and may provide or require the use of company templates or forms, acceptance criteria and approaches to addressing failures or non-conformances.

(2) external requirements such as those relating to ISO and ASTM standards- product related standards may specify target nominals or acceptable ranges/tolerances for product specifications. Compliance to state of the art and generally accepted state of the art is a requirement for many notified bodies and competent authorities.

(3) the type of test to be completed – visual inspection, manual contact measurement, non-contact (automated) measurement system, destructive tests and so on- the criticality of the parameter to be measured in relation to the product may determine or inform the test method format. Manual inspection or test methods may not provide data with the right resolution. While automated systems can be expensive and can take time and resources to quality and validate their use.

(5) Product range and specifications- the product specifications is a critical input as the accuracy, range and resolution of equipment needs to suitable for use in respect of the specification nominal and tolerances.

Example 1

A packaging company has a seal strength on the lid of a package. It wants to put in place a test method to test the seal strength of the package. This scenario would call for a test method validation.

Example 2

A medical device incorporates the use of a spring that is used to actuate a valve. The manufacturer of the device wants to develop a test method that examines the tensile strength of the spring on an ongoing basis. This scenario would also call for a test method validation.

Example 3

A contact lens manufacturer uses an optical comparator to measure the diameter of contact lenses during manufacture. The manufacturer must develop and validate a test method to facilitate the measuring of contact lenses.

Definitions

Attribute: is defined as the result of a property or characteristic. It is generally used with the terms pass or fail.

Accuracy: can also be defined as trueness. An expression of the closeness of agreement between the value that is accepted, either as a conventional true value or an accepted reference value and the value obtained. A system with low bias implies good accuracy and vice versa.

ANOVA (Analysis of Variance): a statistical method used to evaluate the significance of differences in means due to different factor-level combinations.

Bias: The difference between observed "average of measurements" and a reference value; also referred to as accuracy.

CQA (Critical-to-Quality): a property or characteristic with specific nominal value and appropriate limit and range providing a particular quality attribute. A CQA typically is classed as a high risk requirement, where the safety or efficacy of the product depends on the CQA been within the specified limits

Critical Process Parameter (CPP): a process parameter that has a direct impact on critical quality attributes.

Dichotomous Variable: an output with only two possible values. Also known as dummy or indicator variable.

Equipment Qualification: establishing documented evidence that the process equipment is suitable for the intended use and is capable of consistently operating within established limits and tolerances under normal operating conditions.

Process Validation: process validation is defined as confirmation via documented evidence that a particular process performs consistently to a high degree of assurance in accordance with predetermined specifications under anticipated conditions.

Measurement Capability Index (MCI): the Measurement Capability Index (MCI) represents the capability of the measurement system. It is used to evaluate the capability of the gauge to classify product against predetermined specifications.

MSA: a study to determine the degree of error involved in measuring the given parameter. The measurement system involves the combination of operations, procedures, gauges, instruments, environmental conditions, people and software.

Precision: the degree of agreement (scatter) between a series of measurements when a method is applied repeatedly to multiple samplings of a homogeneous sample or artificially prepared sample under the prescribed conditions. There are three types of precision; repeatability, intermediate precision and reproducibility.

Range: range is defined as the interval between the upper and lower measurements required. The minimum specified range should be within the equipment range and validated to operate at all points within the range.

Ruggedness (Intermediate Precision): variation on different days or with different analysts and equipment. The extent to which intermediate precision should be established depends on the circumstances under which the method is intended to be used.

Resolution: the smallest unit of measure that can be obtained reliably from a measurement device, also known as gauge discrimination.

Gauge R&R: represents the estimate of the measurement variation. The measurement variation has two components; repeatability or the precision under the same operating conditions (same operator, test method, sample, etc.) and reproducibility or the precision between operators when measuring the same sample with the same gauge.

Variable: is generally the output that is measured.

Validation: confirmation by examination and provision of objective evidence that the particular requirements for a specific intended use can be consistently fulfilled.

Scenarios

New Test Methods

A test method procedure should be created as early on as possible and trialled and examined for completeness and appropriateness. If new test methods are required, a revision controlled draft should be available for the purposes of the test method validation.

Changes to Existing Methods

If changes to existing test methods are required, a redlined version highlighting the changes should be made available for the test method validation.

Method Transfer

If an existing test method is suitable for the test method validation, a suitability report can be completed to document the suitability and show that all factors have been considered (see attachment 1). However, the test method should have been previously validated. The parameters at which the validation is to be conducted must be within the existing validated range.

Equipment used in a test method must be assessed to ensure the process is within the equipment qualification. All validation testing must be done on qualified equipment. Equipment qualification is therefore a prerequisite of test method validation.

Ruggedness

Ruggedness refers to the variation, on different days or with different operators or equipment. The extent to which ruggedness (aka intermediate precision) should be established depends on the circumstances under which the method is intended to be used.

An initial ruggedness assessment should be completed to understand the sources of variation. More formal ruggedness studies may be required which should be captured in a formal study protocol.

The output of any ruggedness studies should detail any changes or modifications to the test method procedure.

Generally, a scoring system is used to describe ruggedness which forms a ruggedness assessment. As a result of ruggedness studies and consequent updates to the procedure, the ruggedness assessment needs to be reassessed. This reassessment will be reflected in the final scores of the Ruggedness Assessment Matrix.

Key Concepts

Accuracy

Accuracy is a measure of exactness of the test method output or another way of putting it is the closeness of agreement between a set of test results.

For example, take a component that weighs exactly 4 kg according to an NIST traceable scale. If the weight of component is taken 10 times on the balance under study using the test method under study then calculate the mean weight of the 10 readings. The offset between the mean weight and the 4kg "accepted reference value" is a measure of bias.

A large bias = poor accuracy. A small bias = good accuracy.

It is important to note that accuracy does not address the variation between individual measurements.

Simply put, if the average is very close to 4kg, then the test method could have been declared to be very accurate.

It is advised that you consult any relevant standards (e.g. ISO, ASTM) to the product or feature being measured as standards often will call out an accuracy requirement. Generally, results should be accurate to ±1% of the measured value. Therefore, the equipment must be fit for the intended purpose or the measurements in mind.

Note: instrument or equipment accuracy can normally be found on calibration certs provided by the manufacturer or vendor.

Precision

The precision of a method is the degree of agreement among individual test results when the same test method or procedure is applied repeatedly to multiple samplings that represent a population. Precision can be a measure of either the degree of reproducibility or of repeatability of the method. Repeatability refers to the use of a method using the same operator/test person with the same equipment. Repeatability should be assessed using either a minimum of 9 determinations covering the specified range for the method (e.g. 3 concentrations/3 replicates each). Reproducibility refers to the use of the analytical method in different laboratories such as in a collaborative study.

Ruggedness

Intermediate precision (also known as ruggedness) expresses differences related to laboratory variation, as on different days, or with different analysts or equipment within the same laboratory. The extent to which intermediate precision should be established depends on the circumstances under which the method is intended to be used. The effects of random events on the precision of the analytical method should be established. The use of experimental design (matrix) may be used to study the effects of typical variation (dominance factors) on the analytical method (e.g. equipment, analyst, days).

Representative/Continuous Sampling

Representative sampling is used to determine overall process performance (e.g. Pp / Ppk), which is more applicable for processes known or suspected as less than stable or not in statistical control. Sampling in this way best determines overall spread, which includes within-time and time-to-time variation.

Below, some examples are given on how to sample representatively:

(1) Sampling over a given time-period: e.g. a tray of product is produced every 15 minutes, the period of interest is a 1 hour interval and the sample size is 40.

(2) Sampling a batch or product lot not assembled in any order: if the product is packed in a tray (without any grouping) then sample from various sections of the tray.

Consecutive Sampling

This type of sampling involves taking one sample immediately after each another for the subgroup or time period in question, and is used to determine process capability (e.g. Cp / Cpk). Consecutive sampling is used in particular to create control charts where a process is sampled in time order by selecting a subgroup sample consecutively and repeating this sampling over a number of subgroups while in same time order. This method is typically used when the process is stable as there will be little or no causes of lot-to-lot variation.

Range

The range is defined as the interval between the upper and lower measurements required. The minimum specified range should be within the equipment range and validated to operate at all points within the range. If an existing test method or piece of equipment is to be used, it is important to determine if the method parameters for the new/modified test method are within the validated range of the equipment qualification. Remember, all validation testing must be done on qualified equipment. Typically, the equipment qualification assessment is documented in the test method validation protocol.

Resolution

We have previously defined resolution as the smallest unit of measure that can be obtained reliably from a measurement device or system. For example, a Vernier callipers may have different models with different resolutions. Some will have only two digits to the right of the decimal point (X.XX mm) and other models could read three digits to the right of the decimal point (X.XXX mm). The instrument resolution should be better than the resolution of the product specification. If the product specification is X.XXX, then at least a "four-digit" measurement device should be used.

Probability of False Alarms P (FA)
This signifies the likelihood of rejecting a conforming unit. This is typically an acceptance criterion for attribute tests. Refer to MSA template for further illustration.

Probability of Misses P (M)
This indicates the likelihood of accepting a non-conforming unit. This also is typically an acceptance criterion for attribute tests. Refer to MSA template for further illustration.

Protocols
Typically, an approved template is used to create a validation protocol. The protocol sets out the approach to the validation i.e. the approach to qualify the test method. Refer to the appendix for an example of a validation protocol template.

Attachments to the protocol should include ruggedness assessments completed and references to supporting studies/reports. The drafted or "redlined" test method should be attached to the protocol also. The type of MSA protocol (attribute or variable) should also be determined in the validation protocol.

Test Method Accuracy
Accuracy is influenced by both the instrument (scale) and the test method. If you drop the object on the scale and take a reading before the scale has stabilised, the accuracy is likely to be poorer than when using a test method that demands allowing the scale to stabilise. Examples include: Tensile strength at break - strength does not exist as a material property independent of the test method used to measure it.

For properties like time, distance, and mass, there are NIST traceable standards that can be measured. These standards have a generally accepted reference value that can be compared to the observed readings to assess accuracy (bias). No such reference sample exists for tensile strength at break, deflation time or implant radial strength. For tests without a reference value, the accuracy of the underlying sensor (e.g. load cell) used to determine the output should be addressed if possible.

MSA Studies
A measurement system analysis (MSA) is an experimental design used to identify the elements that affect measurement variation. There are two types of data in which MSA studies can be completed i.e. variable data and attribute data. These terms are defined below.

> **Variable data:** data that can assume a range of numerical responses on a continuous scale. Most measurements yield variable data.
>
> **Attribute data:** data that represents the absence or presence of a characteristic.

Non-destructive tests: test where the measured characteristic is not altered due to testing. Since the sample is not altered, multiple readings can be taken on the sample with the expectation of getting the same measured result. Destructive tests: test where the measured characteristic is changed due to testing. Since the sample is changed, there is no expectation of getting the same measured result over multiple readings. The four types of MSA studies include:

- Variable / Non-Destructive
- Variable / Destructive
- Attribute / Non-Destructive
- Attribute / Destructive Table

General MSA Requirements:

Test Environment Conditions - the test environment (i.e. temperature, humidity) should represent the conditions going forward. The effect of multiple environmental conditions can be evaluated if the study is properly designed and planned.

Sample Range - samples should cover the expected range of measurements.

Standard (for attribute MSA) - define the true answer (pass or fail). The standard is based on the inspection ratings of an expert opinion or a measurement system with known better inspection capability than the one under evaluation.

Measurement Instructions/ Training - follow the inspection instructions as defined in the controlled documents or redlines included with the protocol. Do not minimise variability by adding special instructions not defined in the controlled documents or redlines included with the protocol. Reference the controlled documents in the protocol. Special instructions are allowed when using pseudo samples provided that the variability is not minimised due to the instructions. Testers should have a high degree of skill and experience. Do not use new personnel or inexperienced people to conduct measurement studies.

Equipment Qualification and Calibration – The equipment must be calibrated prior to conducting the study. Evidence of the calibrated state should be documented in the report (e.g. calibration certificates etc.). It is important not to re-calibrate the equipment during the study as results can be different due to the calibration effect. The effect of calibration can only be evaluated if the study is properly designed.

Randomisation -

1. Assign the samples to the first operator in random order. Operator measures the parts.
2. Assign the samples to the second operator in random order. Operator measures the parts.
3. Assign the samples to the third operator in random order. Operator measures the parts. Repeat the process described in steps 1 to 3 with the operators for a second and third trial.

Data collection - when documenting the results of a trial, the operator should not have access to the results from the previous trials. A different data collection sheet must be provided for each operator involved in each trial. In lieu of a different data sheet, a data recorder may be used to blind the data recording operator to the test data of previous runs.

Variable MSA Studies :

Non Destructive/Variable MSA Studies

The key requirements for non-destructive and variable MSA studies include:

No. of Operators - at a minimum, 3 operators should be used during the study. More operators are also recommended if human/operator interaction is a source of measurement error.

Sample Size - a minimum of 10 units is recommended.

Trials - a minimum of 3 trials should be completed.

Destructive/Variable MSA Studies

If a test is destructive in nature, repeated measurements cannot be taken as the sample is damaged or destroyed as part of the test. One solution is to adopt standardisation of units where homogeneous samples are created by standardising the material or manufacturing process.

No. of Operators - 3
Sample Size - 10 units
Trials - 3 trials

This equates to 90 measurements in total. If standardisation is not feasible, the use of non-destructive pseudo-samples can be used. However, equivalence should be demonstrated between the pseudo sample and "true" units.

Attribute MSA Studies

Non Destructive
The recommended and minimum sample size requirements for attribute/non-destructive MSA studies are shown below:

Recommended Minimum Sample Size Requirement:
Operators - 3
Sample size - 25
Trials – 3

Destructive
When the test is destructive, repeated measurements cannot be taken as the sample is destroyed or altered. Some approaches are outlined below in order to quantify the measurement variability for destructive tests.

Standardisation Approach: homogeneous and representative samples are created by standardising the method of sample preparation, or material.

Sub-samples: cut each sample into three sub-samples to represent the three trials.

Pseudo-samples: create non-destructive pseudo-samples, documenting a rationale justifying the equivalence of the pseudo samples to the true samples.

Measurement Capability Index
The Measurement Capability Index (MCI) is calculated to assess the capability of the measurement system. The MCI is calculated as a % tolerance.

Measurement Capability Index acceptance criteria:

This index is used to evaluate the capability of the gauge to classify product against the specifications.

The index represents the % of the tolerance (upper specification limit (USL) and the lower specification limit (LSL) that is consumed by the measurement system variation.

Risk Management for Product Lifecyle

This section reviews the themes within ISO 14971, Risk management system. The standard provides a framework on how to apply risk management to medical devices. It sets out the requirements on development, implementation and maintenance of a risk management process for medical devices. It is acknowledged as the principal standard to use when conducting medical device risk management activities. Risk is the combination of the probability of occurrence of harm and the severity of that harm Source – ISO/IEC Guide 51:1999).The term "risk" within the scope of the ISO 14971 International Standard on refers to safety or performance requirements of the medical device or meeting applicable regulatory requirements.

Components of risk

Risk has two main components, that are to be dealt with independently and separately. Firstly, the probability of occurrence of harm and secondly, the severity of that harm.

The various risks presented by a particular device depends substantially on its intended purpose and the effectiveness of the risk management techniques used in the design, manufacture and subsequent use by the end user. Principles of risk management are best applied using a Process and Iterative Approach. A process works to ensure requirements are documents, instructions and templates are in place and roles and responsibilities are defined. An effective risk management process will often have many work instructions or SOPs providing the requirements for aspects of risk management such as PFMEAs, risk planning, risk review, post marketing surveillance and so on.

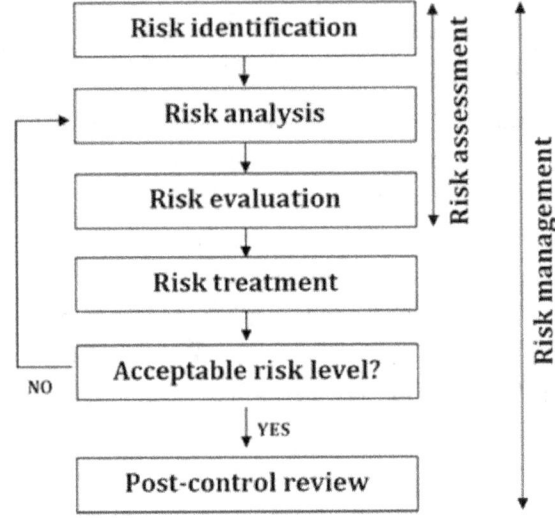

Risk Management Process

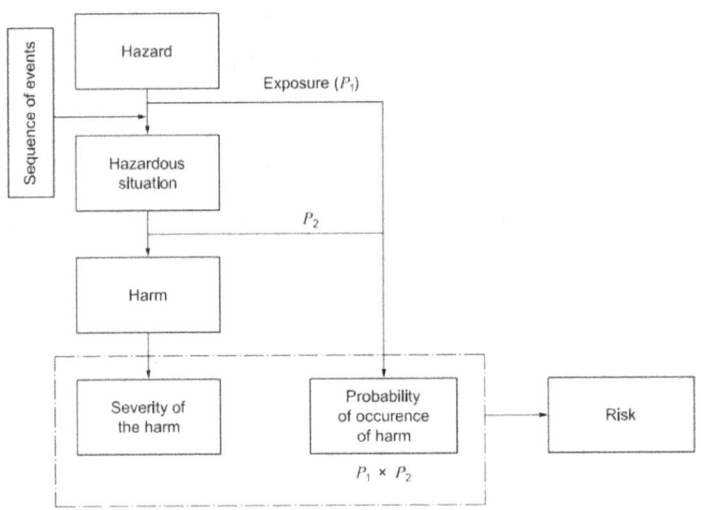

P1, the probability of a hazardous situation occurring.

P2, the probability of a hazardous situation leading to harm

Basic Terms and Definitions

Harm

Harm occurs when ac physical injury or damage to the health of people, or damage to property or the environment.

Hazard

A hazard is a potential source of harm.

Hazardous situation

circumstance in which people, property or the environment are exposed to one or more harms.

Risk

combination of the probability of occurrence of harm and the severity of that harm

Risk Analysis

The systematic use of available information to identify hazards and to estimate the risk. Risk analysis also refers to the analysis of the various sequences of events that can produce hazardous situations and harm

Risk Control

process in which decisions are made and measures implemented by which risks are reduced

Risk Estimation

A process used to assign values to the probability of occurrence of harm and the severity of that harm

Risk Management

systematic application of management policies, procedures and practices to the tasks of analysing, evaluating, controlling and monitoring risk

Safety

freedom from unacceptable risk

Severity

the measure of the possible consequences of a hazard

General Requirements

General requirements for risk management system are covered in Clause 4 under sub-clause 4.1 Risk management process. The manufacturer shall establish, implement, document and maintain an ongoing process for:

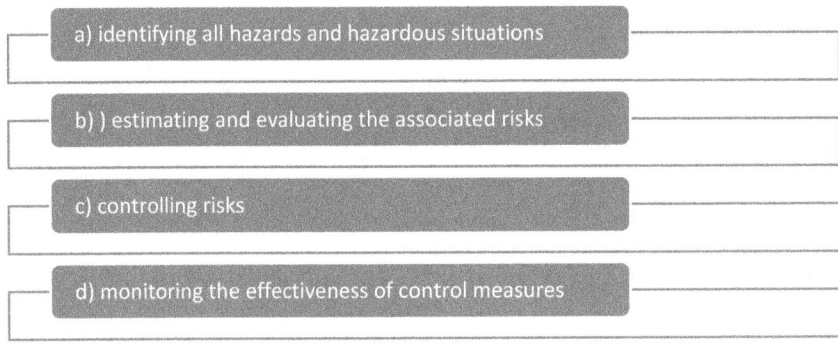

a) identifying all hazards and hazardous situations

b)) estimating and evaluating the associated risks

c) controlling risks

d) monitoring the effectiveness of control measures

Requirements of the manufacturer

In addition, the following ground rules apply:

-The Risk management applies throughout the life cycle of the medical device.

-The process shall include the following elements:

- -RISK ANALYSIS
- -RISK EVALUATION
- -RISK CONTROL
- -PRODUCTION AND POST-PRODUCTION ACTIVITIES

These areas are discussed in further detail further on in this book.

ISO 13485, Quality Management System for Medical Devices and Risk Management

Risk and the application of risk management is also specified in ISO 13485-Quality Management System for Medical Devices, in a number of instances. Specifically, risk management in a component of product realization (7.1) and under the control of design and development. Under the planning of product realization:

> *"The organization shall plan and develop the processes needed for product realization. Planning of product realization shall be consistent with the requirements of the other processes of the quality management system. The organization shall document one or more processes for risk management in product realization. Records of risk management activities shall be maintained."* Reference, ISO 13485

Under ISO 13485 section 8.2, Monitoring and measurement, 8.2.1 Feedback the information provided from monitoring activities (complaints, production data and post-production data) is suggested as potential input to risk management. In turn, this should inform the product design and development process (product realization) and general improvement.

2017/745 (EU MDR) & Risk Management

MDR specifies that Devices shall achieve *"performance intended by their manufacturer and shall be designed and manufactured in such a way that, during normal conditions of use, they are suitable for their intended purpose."*

They shall be safe and effective and shall not compromise the clinical condition or the safety of patients, or the safety and health of users or, where applicable, other persons,

Any risks which may be associated with their use must result or constitute in *"acceptable risks when weighed against the benefits to the patient and are compatible with a high level of protection of health and safety, taking into account the generally acknowledged state of the art."*

Under the general safety and performance requirements (GSPR) device risk must be as low a possible and give due diligence to residual risks or in other words risks that remain after all possible risk reduction, control and mitigation. GSPR Annex 1 Section 2 and 3 details the requirements on the risk benefit ration and on risk management and the lifecycle or risk management applied to products.

GSPR Annex I Section 3 of MDR 2017/745, Annex I

The requirement in this Annex to reduce risks as far as possible means the reduction of risks as far as possible without adversely affecting the benefit-risk ratio.

3. Manufacturers shall establish, implement, document and maintain a risk management system. Risk management shall be understood as a continuous iterative process throughout the entire lifecycle of a device, requiring regular systematic updating. In carrying out risk management manufacturers shall:

(a) establish and document a risk management plan for each device;

(b) identify and analyse the known and foreseeable hazards associated with each device;

(c) estimate and evaluate the risks associated with, and occurring during, the intended use and during reasonably foreseeable misuse;

(d) eliminate or control the risks referred to in point (c) in accordance with the requirements of Section 4;

(e) evaluate the impact of information from the production phase and, in particular, from the post-market surveillance system, on hazards and the frequency of occurrence thereof, on estimates of their associated risks, as well as on the overall risk, benefit-risk ratio and risk acceptability; and

(f) based on the evaluation of the impact of the information referred to in point (e), if necessary amend control measures in line with the requirements of Section 4. 4. Risk control measures adopted by manufacturers for the design and manufacture of the devices shall conform to safety principles, taking account of the generally acknowledged state of the art.

To reduce risks, Manufacturers shall manage risks so that the residual risk associated with each hazard as well as the overall residual risk is judged acceptable. In selecting the most appropriate solutions, manufacturers shall, in the following order of priority:

(a) eliminate or reduce risks as far as possible through safe design and manufacture; where appropriate, take adequate protection measures, including alarms if necessary, in relation to risks that cannot be eliminated; and (c) provide information for safety (warnings/precautions/contra-indications) and, where appropriate, training to users.

Manufacturers shall inform users of any residual risks.

GSPR Annex I MDR 2017/745, Section 5 States

In eliminating or reducing risks related to use error, the manufacturer shall:

(a) reduce as far as possible the risks related to the ergonomic features of the device and the environment in which the device is intended to be used (design for patient safety), and
(b) give consideration to the technical knowledge, experience, education, training and use environment, where applicable, and the medical and physical conditions of intended users (design for lay, professional, disabled or other users).

GSPR Annex I MDR 2017/745 Section 6

The characteristics and performance of a device shall not be adversely affected to such a degree that the health or safety of the patient or the user and, where applicable, of other persons are compromised during the lifetime of the device, as indicated by the manufacturer, when the device is subjected to the stresses which can occur during normal conditions of use and has been properly maintained in accordance with the manufacturer's instructions.

Devices shall be designed, manufactured and packaged in such a way that their characteristics and performance during their intended use are not adversely affected during transport and storage, for example, through fluctuations of temperature and humidity, taking account of the instructions and information provided by the manufacturer.

All known and foreseeable risks, and any undesirable side-effects, shall be minimized and be acceptable when weighed against the evaluated benefits to the patient and/or user arising from the achieved performance of the device during normal conditions of use.

For the devices referred to in Annex XVI, the general safety requirements set out in Sections 1 and 8 shall be understood to mean that the device, when used under the conditions and for the purposes intended, does not present a risk at all or presents a risk that is no more than the maximum acceptable risk related to the product's use which is consistent with a high level of protection for the safety and health of persons.

Risk Analysis

Risk analysis process can be sub divided into 4 parts which are detailed in ISO 14971. These include:

RISK ANALYSIS

(1) description of the intended-use of the medical device and reasonably foreseeable misuses
(2) Identification of the characteristics of the medical device that are related to safety
(3) Identification of hazards and hazardous situations associated with the medical device
(4) estimation of risk for each hazardous situation

Understanding the Intended use of a medical device is fundamental as it determines the proper application and use of the device. Designers aim to properly define the intended use as it then allows them to focus on what specific requirements will deliver such a device, meeting the user requirements and intended use. Intended is concerned with (1) medical indication, (2) patient population, (3) user profile (e.g. doctor or lay person) (4) part of the body or tissue the device is concerned with and (5) the use environment.

Reasonably foreseeable misuse

This is a scenario when a medical device is used in a way that it was not intended or designed to be used as set out by the manufacturer. Situations of reasonably foreseeable misuse are understood as situations that can anticipated based on human behavior. (hence reasonably foreseeable).

As part of risk management reasonably foreseeable misuses should be identified by the manufacturer. These can be identified in a number of ways which include:

- During product realization and Design and Development)
- Simulated studies such as Usability Engineering Studies
- Customer Complaints or adverse events during Post market monitoring.

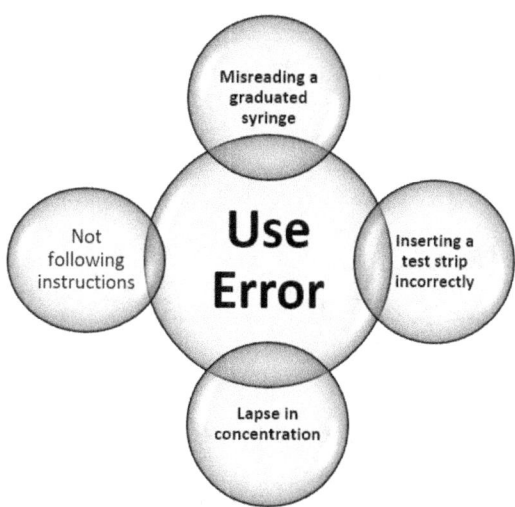

Identification of characteristics related to safety

Certain characteristics of a medical device can affect safety! It is up to the manufacturer to identify the performance requirements or the functions of the medical devices that fulfil the intended use or if hazardous situations can occur that may impact the performance of safety. Refer to section on Risk Analysis for information on questions used to Identify hazards and characteristics related to safety

Identification of hazards and hazardous situations

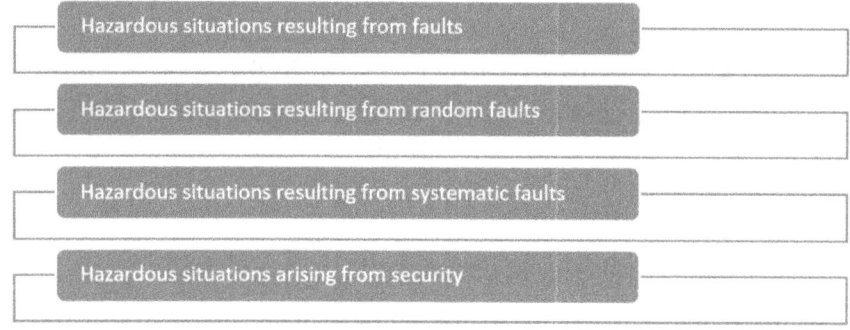

A hazard = potential source of a harm

Hazards can be identified from both the intended use and also any reasonably foreseeable misuse. As previously mentioned, Annex A of ISO 14971:2019 provides details on the characteristics relating to safety. In turn, these characteristics can help in identifying hazards and hazardous situations. Note: Annex C of ISO 14971:2019 provides guidance that can help in identifying hazards and sequences of events that can lead to hazardous situations.

Hazardous Situations

Hazardous situations resulting from faults

If a hazardous situation occurs due to a fault condition, the probability of a fault occurring is not the same as the probability of the occurrence of harm. A fault condition may initiate a sequence of events, however this may not necessarily, result in a hazardous situation. Therefore, a hazardous situation does not result in harm always.

Hazardous situations resulting from random faults

Random faults can be a result physical or chemical corrosion, contamination and mechanical wear-out.

Hazardous situations resulting from systematic faults

The term systematic fault intends to describe when a series of actions/environmental conditions or inputs combine to cause a fault condition. It can be caused by an error in any activity but normally remains latent unless the combination of conditions lead to the fault happening.

Identification of hazards and hazardous situations

This section summaries the questions used in the identification of hazardous situations. (Blue Graphic). For each question, a practical insight is provided to assist in the application of the questions. A common practice of manufacturers is to list the questions and responses in risk documented such as a risk analysis document or a design risk analysis.

Failure Mode and Effects Analysis (FMEA)

The FMEA methodology looks consequences of an individual failure modes that are identified and evaluated based on the process steps (PFMEA) or the component level (DFMEA). FMEAs looking at the manufacturing process or assembly are known as Process FMEAs. If it related to the Design or in which way the device could fail, it is deemed a Design FMEA. If the FMEA is focused on the use or foreseeable misuse of the device it is referred to as a Use FMEA.

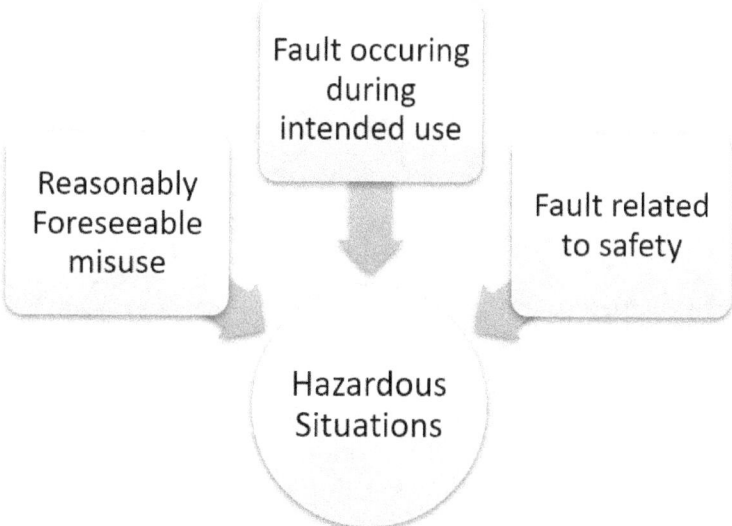

Risk Estimation / Evaluation

Probability

Qualitative estimation of probability is where descriptions are used to estimate probability, High (often), medium (sometimes) low (rare, unlikely). A Quantitative approach is where information or data is used to estimate probability, for example, parts per million. The quantitative approach, due to the fact it is normally based on data can provide more thorough risk estimation and evaluation.

Risk estimation

Risk estimation involves the analysis of the probability of occurrence of a harm and the severity of the harm.

"For each identified hazardous situation, the manufacturer shall estimate the associated risk (s) using available information/data."

Risk estimation incorporates an analysis of (1) probability of occurrence of harm and the (2) severity of the harm.

The qualitative and quantitative systems used for categorization of probability of occurrence of harm and severity of harm need to be recorded in the risk management file to comply with ISO 14971.

High- likely to happen, often or frequently, always

Medium- can happen, but not frequently

Low -unlikely to happen, rare and remote

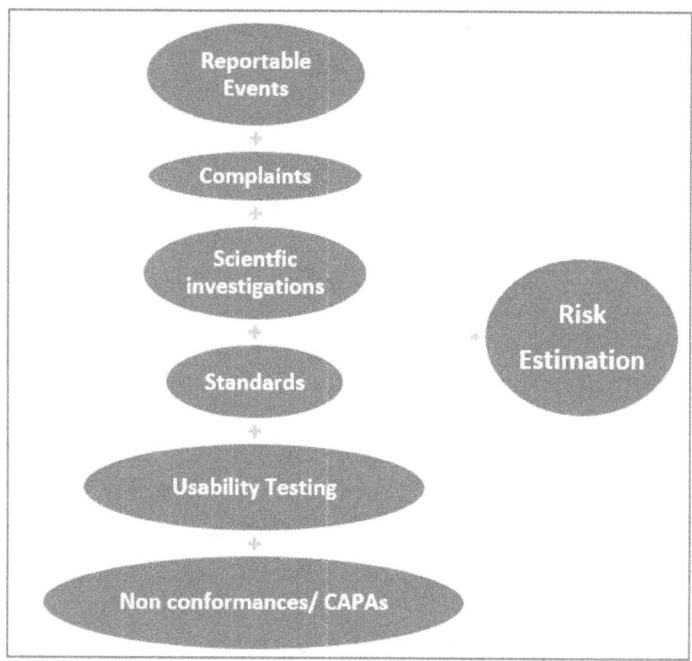

Sources of data for estimation of risk

Risk Control

If risk is too high then further risk control measures must be implemented to reduce the risk to an acceptable level. A number of actions can be taken in order to further reduce risk including: (1) changing the design to reduce risk- safety that is built-in or inherent in the design is very effective. (2) introducing protective measures in the device or the manufacturing process, (3) including a warning statement into the instructions for use (IFU). (4) Use of symbols on labelling and cartoning- information on safety if communicated and used by the user can provide mitigation also.

Risk Acceptability

ISO 14971 requires companies to document a Policy and develop criteria for risk acceptability. The policy provides instruction on how to establish the criteria for acceptability of the overall residual risk. It should address individual residual risks and the risk-benefit ratio or analysis also.

ISO 14971 requires a policy for establishing the criteria for risk acceptability. The policy can include (1) purpose, (2) scope, (3) factors and considerations for determining acceptable risk, (4) approaches to risk control and (5) requirements for approval and review.

- The purpose should detail the specific goals of the policy for establishing criteria for risk acceptance/acceptability.

- Factors and considerations for risk acceptability- Applicable regulatory requirements and international standards for the medical device,

Criteria for risk acceptability

Criteria for risk acceptability should be established in advance of any risk management activity or execution of the risk management plan so that guidance is available in determining acceptable risk. The policy is normally is included in a risk management procedures or other quality document.

Evaluation of overall residual risk and acceptability

An evaluation of the overall residual risk and acceptability must be completed in accordance with the risk policy. ISO 14971 specifies that both the method and the criteria be stated in the risk management plan.

Clause 8:

"Evaluation of overall residual risk After all risk control measures have been implemented and verified, the manufacturer shall evaluate the overall residual risk posed by the medical device, taking into account the contributions of all residual risks, in relation to the benefits of the intended use, using the method and the criteria for acceptability of the overall residual risk defined in the risk management plan. If the overall residual risk is judged acceptable, the manufacturer shall inform users of significant residual risks and shall include the necessary information in the accompanying documentation in order to disclose those residual risks.

If the overall residual risk is not judged acceptable in relation to the benefits of the intended use, the manufacturer may consider implementing additional risk control measures or modifying the medical device or its intended use. Otherwise, the overall residual risk remains unacceptable. The results of the evaluation of the overall residual risk shall be recorded in the risk management file. Compliance is checked by inspection of the risk management file and the accompanying documentation"

Risk management Review and Reporting

The risk management review is an important step before the commercial release of the medical device and required per ISO 13485, clause 9, The final results of the risk management process, as obtained by executing the risk management plan, are reviewed. The risk management report contains the results of this review and is a crucial part of the risk management file.

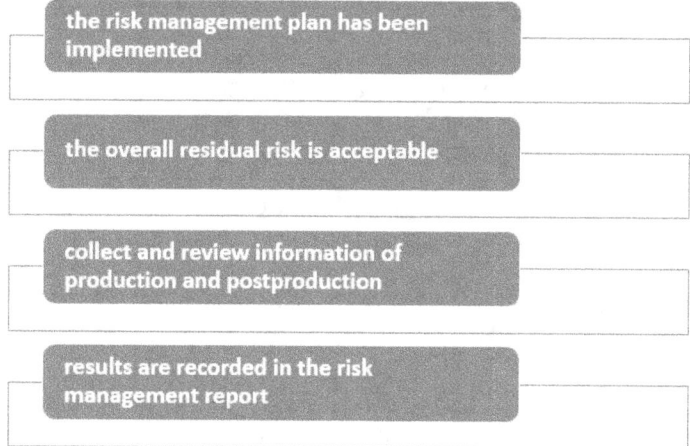

- the risk management plan has been implemented
- the overall residual risk is acceptable
- collect and review information of production and postproduction
- results are recorded in the risk management report

The report serves as the high-level document that provides evidence that the manufacturer has ensured that the risk management plan has been satisfactorily fulfilled and the results confirm that the required objective has been achieved. Subsequent reviews of the execution of the risk management plan and updates of the risk management report can be needed during the life cycle of the medical device, as a result of the execution of production and post-production activities.

When
- The risk management review should be completed when the implementation of the plan is completed along with verification of the risk control measures. This should occur in advance of product release of commercial product.

What
- The risk management report is the conclusing output of all risk management activities.

Schedule

- The risk management file must be maintained over the life cycle of the product. The risk mangament report should therefore be reviewed at appropriate intervals to ensure it is accurate and current. Other scenarios that require review or update of the risk management report would be a change in the maufacturing process that would require re-validation of the process or a new product type been added to a current range

Effectiveness

- The risk management process must be suitable, fit for purpose and demonstrated as effective.

Severity

Severity is based on the harm that could result from the use of a particular medical device. Severity must be included and documented in the Risk Management File in accordance with the requirements of ISO 14971:2019.

Risk Management File

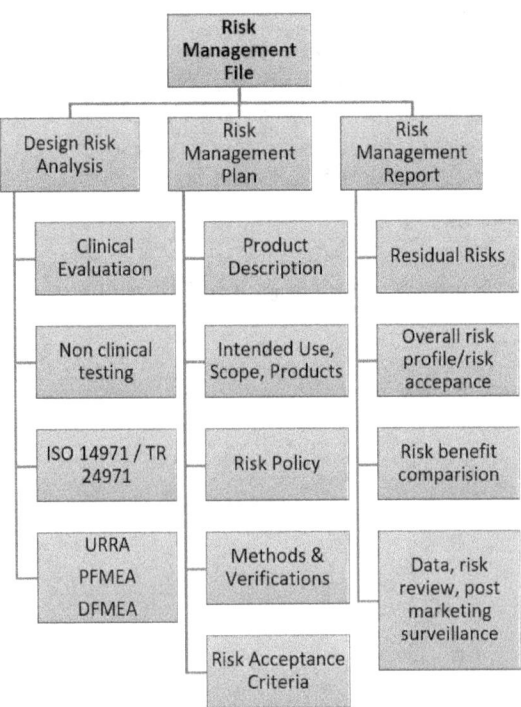

Overall Residual Risk

Benefit-risk analysis

The purpose of a Benefit-risk analysis is to document that residual risks are outweighed by the benefits of the device. When conducting the benefit-risk analysis, the criteria for acceptable must be taken from the risk management plan. Risks that are not acceptable and where additional risk control is not possible, the benefit-risk analysis must conclude that the benefits outweigh the residual risks.

Criteria of benefit-risk analysis

The benefit-risk analysis should take into regulatory factors but also the clinical and medical outcomes as a result of the availability of a device or product.

Hence, prior to commercial launch, clinical investigations may be required to determine that the balance between benefit and residual risk is acceptable and that the product is safe and effective with acceptable probabilities of occurrence of harm. Benefit risk analysis can be aided by comparative review of similar devices that are already on the market and review of generally acknowledged state of the art principles.

Residual Risk

After risk control and risk mitigation measures are applied, a new (follow up) risk assessment should completed to determine residual risks. If the residual risk is deemed unacceptable then further risk control measures should be applied.

Post Production Review

The goal of post-production risk management activities is to drive updates to the risk management file based on :

- Unanticipated risks or new or emerging risks identified
- Risks associated with product-use (use errors)
- Product performance levels relative to residual risk levels
- the likelihood of occurrence for hazards/harms change from previous established levels

Based on this information, the Risk Management report should be kept under review and updated if the risk profile is changed.

FMEA, Failure Mode and Effects Analysis

Failure Modes & Effects Analysis (FMEA) is a risk management analysis tool used during the design or to assess the manufacture of products to that can be used to manage risks caused by failure modes.

Types of FMEA

UFMEA assess the failure modes that occur during product-use and examines the robustness of

product design, the intended use and also any reasonably foreseeable misuse by the end-user.

DFMEA assess the failure modes related to design of a system, product,

feature, component of a final product and degradation of the product

over its expected life.

PFMEA assess the failure modes that are related to the manufacture

process of the product, including the safety of process operators.

Role of Standards

International standards such as ISO, ASTM, IEC are commonly used alongside ISO 14971 in regards to medical devices. These standards provide designers and manufacturers with information that has been reviewed by working groups and experts relevant to their industry. There publication is gated by peer evaluation, drafts that are subject to review and feedback and eventually subject to voting.

When performing risk management, the manufacturer first considers the medical device being designed, its intended use, its characteristics related to safety, and the associated hazards and hazardous situations.

Designers and Manufacturers can identify the product standards and process standards that contain specific requirements and help reduce the risks of use.

There are two distinct types of standards. (1) Product standards and Process standards. Both types play a part in delivering safe and effective medical devices, however, the product standards are more specific to the device in its scope.

Product standards can utilize the following:

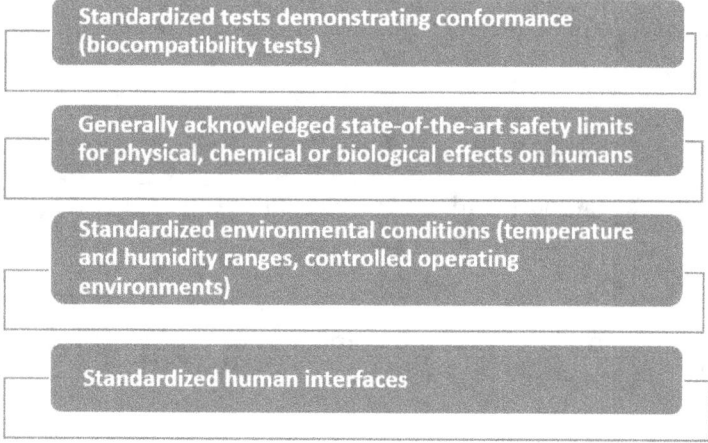

Process standards can be used in the following ways:

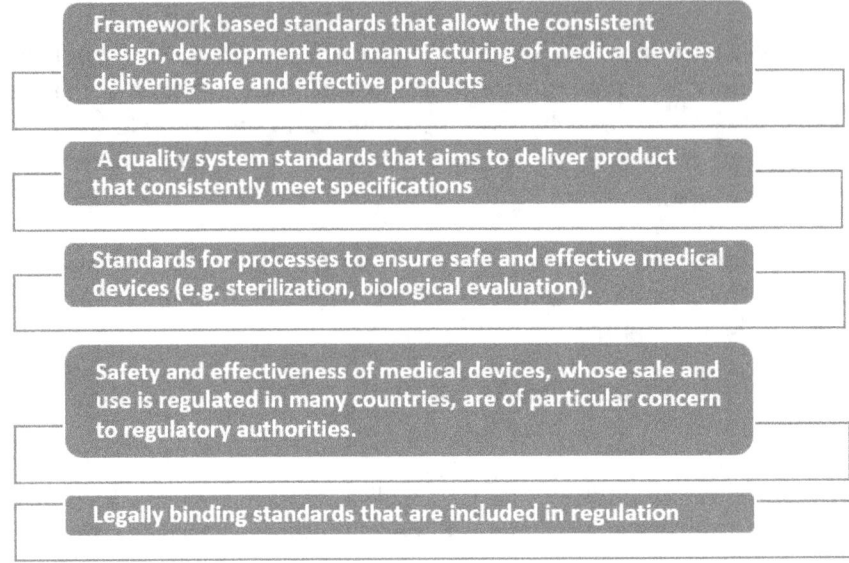

The following are examples of process standards applicable to manufacturers that can be used to assist in the Risk Management activities within a company:

ISO 16142-1 Essential Principles relating to Risk

This is a product safety standard and addresses safety and performance requirements that should be considered. Specifying the correct safety and performance requirements can result in successful risk reduction and risk acceptability.

ISO/IEC Guide 63, Guide to the development and inclusion of aspects of safety in International Standards for medical devices

This guide provides recommendations on the development and inclusion of safety requirements for international standards for medical devices. Their uses can be summarized as follows:

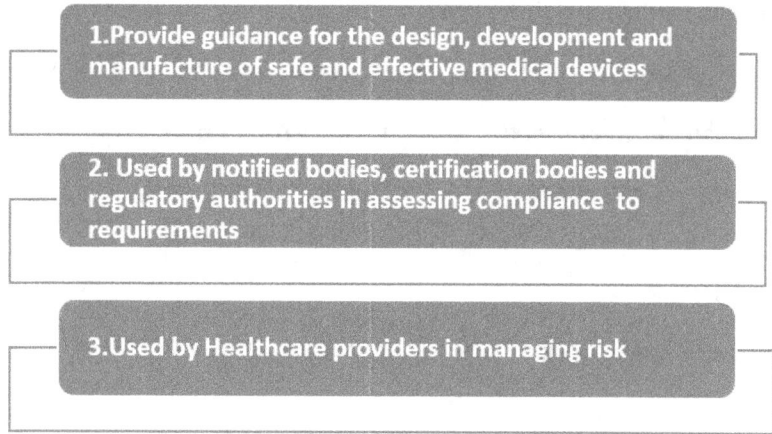

IEC 62366-1, Medical devices, Application of usability engineering to medical devices

Risk Management has a number of inputs that help inform and assist in the application of the principles of risk management, e.g. identification of harms, hazardous situations- risk estimation/evaluations and risk analysis.

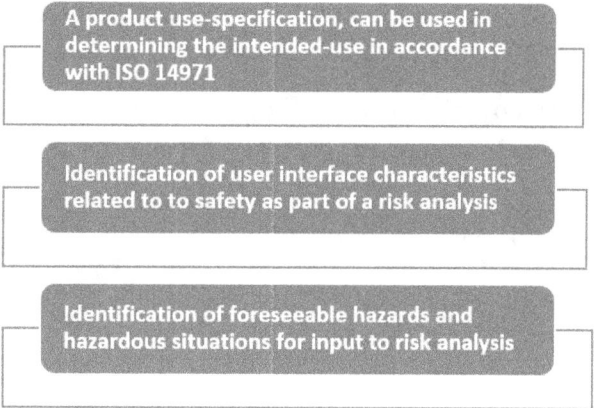

ISO 10993-1, Biological evaluation of medical devices - Part 1: Evaluation and testing within a risk management process

ISO 10993-1 covers the biological evaluation of medical devices within a risk management process, and takes into account the overall evaluation and development of each medical device.

In applies the same approach of ISO 14971 and involves:

    (1) the identification of biological hazards for the medical device

(2) estimation and evaluation of the risks,

(3) the control of risks and monitoring the effectiveness of the risk controls

The biological evaluation utilizes the following information:

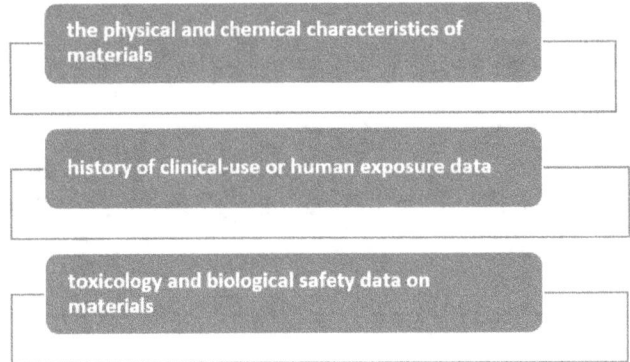

It is important to state that the Biological Safety Evaluation Report, is an element of the risk management file and may be referenced in the risk management report or other risk documentation, as required.

ISO 14155 - Clinical investigation of medical devices for human subjects — Good clinical practice

ISO 14155 is a standard that covers good clinical practice for medical devices for human subjects. It can be of use in determining the clinical risks and the benefit-risk analysis.

Failure Modes And Effects Analysis

Introduction

The Acronym FMEA otherwise known as Failure modes and effects analysis (FMEA) is a methodology of evaluating a product or process to identify the potential ways in which it may potentially fail. During the process of FMEA once the potential failures are identified- then the effects of the mode of failure on safety, efficacy, performance of of the product or indeed the process (or both). FMEA can be rolled out to hardware, software, processes including human action, and their interfaces, in any combination.

Business and Quality Case for FMEA

Why in simple terms FMEA is a risk management risk assessment tool it also can be used to improve business metrics, reduce waste, reduced we rework, improve quality, and from a safety perspective reduce or prevent harm, injury or adverse events from occurring. The process of completing a female start with identifying the process steps. Each process step can be reviewed to identify potential failures that can lead to hazards are hazardous situations. Once these are identified the next step is produce the likelihood of failures, eliminate them if possible by design, and also reduce the effects are the consequences of the failure mode. The cause of the failures can also be documented. Taking a business viewpoint can bring an awareness of the effect off failures on yield, downtime, scrap, cost and customer satisfaction. For quality, and the safety and performance of the finished product, FMEAs for medical devices need to cover the potential harm to patients as a result of hazards and hazardous situations.

Why FMEA?

Methodology for FMEA

There are 3 broad phases of conducting a FMEA which are:

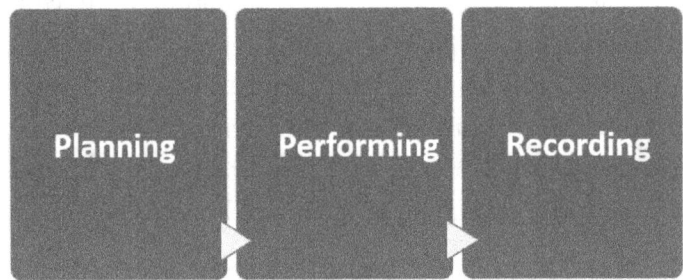

The creation of a FMEA is often an iterative process- meaning it can go through a number of drafts and revisions as it is populated and more information becomes available and risk analysis and estimation is completed The FMEA will, however, always identify the effects of failure modes on the top level of the hierarchy within the analysis scope. In the planning stage, the boundaries of the FMEA should be established. This can be helped by creating a process flow diagram. Some process steps or sub process may be outside the boundary or scope of the FMEA to somewhat make the scope and size of the FMEA manageable or if a separate and independent FMEA is more suited.

Planning an FMEA involves considering why an analysis is to be performed, what equipment or process elements. What is the goal of the FMEA, is it an overarching FMEA or should it focus on each step of a process. After consideration and planning of a FMEA the following information should be understood.

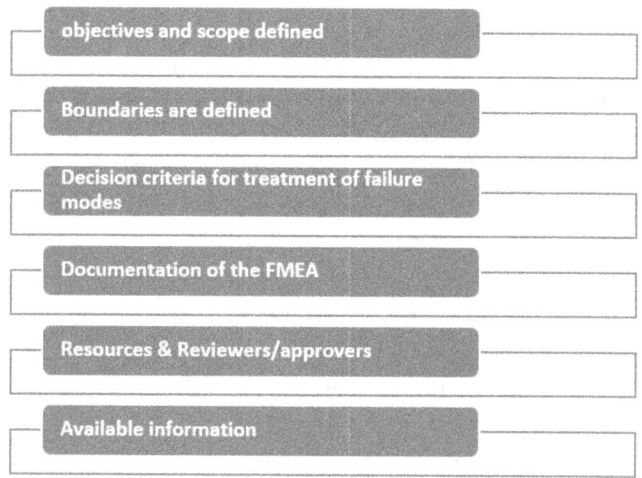

Definition of decision criteria for treatment of failure modes

Decision criteria for the treatment of failure modes helps to focus or prioritise on F.Ms that are the most critical. For medical devices, safety related failure modes, or failure modes that may impact the patient must be treated as critical. A more detailed priority may be established based on the severity of the failure mode. Some failure modes will result in risk that is unacceptable.

Criticality of failure modes can also be assessed by:

- reviewing the severity of the failure effect (on patient, product or process)
- the likelihood that the failure mode might occur and lead the consequence or harm
- detectability of the failure mode

Based on Severity, Occurrence, Detection scores (SXOXD) an RPN, Risk Priority number can be established.

Identify Failure Modes

Failure modes identification:

DFMEA

- If similar devices exist, failure modes may be known or developed based on existing data and knowledge.
- Failure modes may present themselves in different modes of operation
- Failure modes may present themselves at different stages of the product lifecycle
- Can failure modes be a result of storage or transport
- Is there material issues that can result in failure modes

PFMEA

- If a similar or existing manufacturing process is used, the process failure modes may be available
- Is the process subject to environmental changes or trends
- Are there failure modes due to operational issues, human error or automation issues

Identify existing controls or detection methods

After failure modes are identified for each process step or element, the existing control measures and any detection methods should be recorded against each failure mode. Controls may prevent a failure mode or reduce its occurrence while detection allows identification of the failure mode which allows reaction or intervention.

Identify effects of failure modes

A failure effect can be understood as a consequence of a failure mode. Failure effects may be caused by one or more failure modes. The description of each failure effect have a level of detail that allows the assessment of the severity level and what the consequences would be.

Identify failure causes

The cause of the failure and how it occurs is helpful in reducing the likelihood of failure or its consequences.

Common cause and common mode failures

Common cause failures happen when more than one element fails at the same time or within a short period of time that result in the effect of the failures. Elements can also fail in the same way or with the same failure mode- however, this can be due to different causes or the same cause. The likelihood of occurrence can be estimated using:

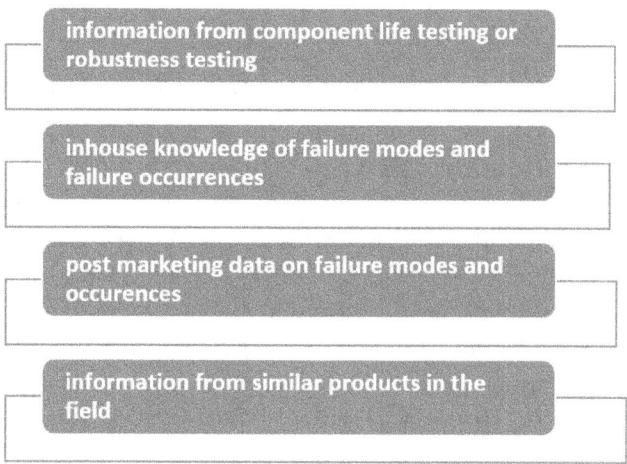

- information from component life testing or robustness testing
- inhouse knowledge of failure modes and failure occurrences
- post marketing data on failure modes and occurences
- information from similar products in the field

Identify actions

This could include additional controls, revalidated, additional audit requirements of engineering changes.

If new controls or detection methods are agreed as necessary and introduced, re-analysis should be completed post implementation of changes. This is to assess if:

- any new failure modes or effects have been introduced (e.g. re-sterilisation may damage packaging ; and

- the criticality of the particular failure modes is deemed acceptable.

ISO 13485 Quality Management System for Medical Devices

ISO 13485 -Medical Devices, Quality Management system is an internal standard for use within medical device manufacturing companies and organisations involved in the design and/or development, production, storage and distribution, installation, or servicing of a medical device and design and development or provision of technical or professional services. It is a standard that applies throughout the product life-cycle.

The recent revision is designed to address recent developments in quality management and other updated regulations that relate to the industry. Improvements in the new version of the standard include broadening its applicability to include all organisations involved in the lifecycle of the product, from the concept stage to end of life along with greater alignment with regulatory requirements and post-market surveillance including complaint handling. ISO 13485:2016 is also used by suppliers or external vendors that provide QMS related services. It has largely become an essential element for manufacturers placing product on the market throughout the world. While it differs to regulations in so far it is an international standard, it has become integrated into many national competent authorities requirements for market approval and certification activities. Requirements within the standard are applicable regardless of the size or an organisations. It should be noted that if clause 6, 7 or 8 of ISO 13485 is not applicable due to the activities undertaken by the organisation or the nature of the medical device for which the quality management system is applied, the organisation does not need to include such a requirement in its quality management system. It uses a process approach to quality management which is made up of subprocesses where inputs are used and outputs inform other processes.

Code of Federal Regulations, FDA, 21 CFR Part 820

It is necessary for manufacturers established a quality (management) system if they intend to market commercial medical device products in the United states. The requirement is concise in nature but it has wide implications for a the manufacturer. In establishing a QMS there needs to be broad commitment within the organization and the resources, training and personnel provided to achieve implementation and in ongoing application and maintenance of the quality management system.

Reference: 820.5 Quality system.

'Each manufacturer shall establish and maintain a quality system that is appropriate for the specific medical device(s) designed or manufactured, and that meets the requirements of this part.'

21 CFR Part 820 consists of 15 subparts:

Subpart A - General Provisions
Subpart B - Quality System Requirements
Subpart C - Design Controls
Subpart D - Document Controls
Subpart E - Purchasing Controls
Subpart F - Identification and Traceability
Subpart G - Production and Process Controls
Subpart H - Acceptance Activities
Subpart I - Nonconforming Product
Subpart J - Corrective and Preventive Action
Subpart K - Labelling and Packaging Control
Subpart L - Handling, Storage, Distribution, and Installation
Subpart M - Records
Subpart N - Servicing
Subpart O - Statistical Techniques

Factors in Device Design, Safety and Performance

The Safety and Performance of medical devices must be a factor from the conception stage of a product through its development, verification validation, manufacturing and to eventual commercialisation. Performance and safety requirements are necessary to deliver products that meet the intended use and provide the desired outcomes for patients and end users. These requirements begin as stakeholder needs. Where upon assessment they manifest themselves as design inputs and design requirements have a product.

The inception of Safety and Performance requirements may seem obvious or expected, however, it goes beyond mere stakeholder needs or good practice as safety and performance requirements ensure patient safety is at the core of design and manufacturing. The application of regulation and appropriate legislation also requires performance and safety is demonstrated prior to approval and through a products lifecycle. Therefore, by placing safety and performance front and foremost, better design processes and outcomes can be achieved during product registration on submission and through the ongoing process of post marketing surveillance and regulatory oversight. 745/2017 (EU) MDR specifies general safety and performance requirements which need to be demonstrated by the manufacturer in relation to products. In addition, competent authorities throughout the world mandate safety and performance requirements as part of legal registration and also per local regulations. Safety and Performance must be accounted for in:

- Both the design activities and manufacturing stages
- Under normal conditions of use

- Safety of the patient but also any users associated with its operation
- Application of risk principles to determine acceptable risks versus the benefit

Safe Medical Devices are in the interest of manufacturers and patients alike as the products must be fundamentally safe in order to provide customer satisfaction and prevent complaints or reporting of harmful effects or events by the patient. 'Safe' can be understood as *freedom from unacceptable risk*. Depending on the classification and intended use of a medical device, it is reasonable that there shall always be a level of risk. This risk however, should be as low as possible and deemed acceptable by experts, regulators and clinicians in accordance with regulations and must provide a clear benefit to the patient or user.

ISO 13485 & Product Realization

Product realization per ISO 13485 is the process of planning product development and introduction but also the subsequent steps that are meaningful in the success of the product introduction. Planning should be initiated early-on in the design stage and should include timelines, resources required, intended markets. One of the most important aspects of planning is gaining the correct stakeholder needs. Again these can be made up of various different inputs. Intended use of the device, will the device be disposable, who is the likely users of the device and so on.

Product realization must establish customer requirements and document the design and development efforts. ISO 13485 also has requirements around purchasing, production, service product and monitoring and measuring equipment. Product realization can be defined as a collection of processes that delivers a product or service to the customer. There is an 'opt out' mechanism that where an organisation can exclude specific requirements, in cases where product realization is not applicable.

Planning of Product Realization / Design and Development Planning

It is the manufacturer's responsibility to establish and maintain plans that describe or reference the design and development activities and define responsibilities for implementation. The plans should identify and describe the interaction with different groups or activities that are part of the design and development process. The maintenance of plans to reflect an accurate state as the design and development progresses is also a key factor. The design and development planning is intended to be prospective in nature. It allows risks to be identified earlier and promotes timely delivery of projects.

Failure Modes And Effects Analysis (FMEA)

Introduction

The Acronym FMEA otherwise known as Failure modes and effects analysis (FMEA) is a methodology of evaluating a product or process to identify the potential ways in which it may potentially fail. During the process of FMEA once the potential failures are identified- then the effects of the mode of failure on safety, efficacy, performance of the product or indeed the process (or both). FMEA can be rolled out to hardware, software, processes including human action, and their interfaces, in any combination.

Business and Quality Case for FMEA

Why in simple terms FMEA is a risk management risk assessment tool it also can be used to improve business metrics, reduce waste, reduced we rework, improve quality, and from a safety perspective reduce or prevent harm, injury or adverse events from occurring.

The process of completing a female start with identifying the process steps. Each process step can be reviewed to identify potential failures that can lead to hazards are hazardous situations. Once these are identified the next step is produce the likelihood of failures, eliminate them if possible by design, and also reduce the effects are the consequences of the failure mode. The cause of the failures can also be documented. Taking a business viewpoint can bring an awareness of the effect off failures on yield, downtime, scrap, cost and customer satisfaction. For quality, and the safety and performance of the finished product, FMEAs for medical devices need to cover the potential harm to patients as a result of hazards and hazardous situations.

Why FMEA?

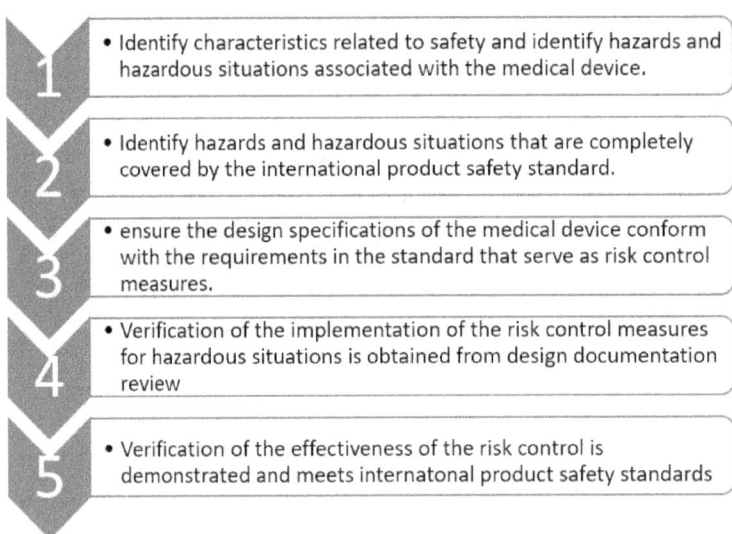

Methodology for FMEA

There are 3 broad phases of conducting a FMEA which are:

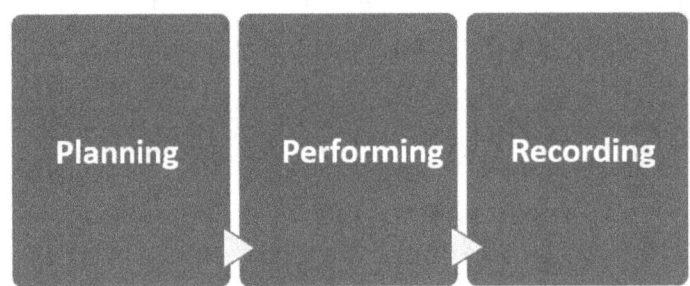

The creation of a FMEA is often an iterative process- meaning it can go through a number of drafts and revisions as it is populated and more information becomes available and risk analysis and estimation is completed. The FMEA will, however, always identify the effects of failure modes on the top level of the hierarchy within the analysis scope.

In the planning stage, the boundaries of the FMEA should be established. This can be helped by creating a process flow diagram. Some process steps or sub process may be outside the boundary or scope of the FMEA to somewhat make the scope and size of the FMEA manageable or if a separate and independent FMEA is more suited. Planning an FMEA involves considering why an analysis is to be performed, what equipment or process elements. What is the goal of the FMEA, is it an overarching FMEA or should it focus on each step of a process. After consideration and planning of a FMEA the following information should be understood.

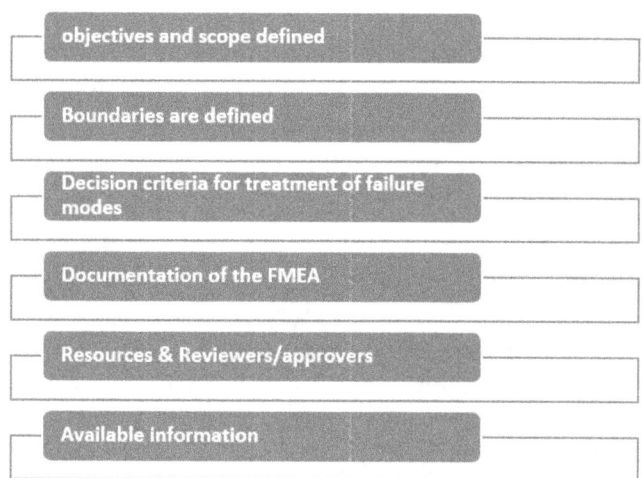

Definition of decision criteria for treatment of failure modes

Decision criteria for the treatment of failure modes helps to focus or prioritise on F.Ms that are the most critical. For medical devices, safety related failure modes, or failure modes that may impact the patient must be treated as critical. A more detailed priority may be established based on the severity of the failure mode. Some failure modes will result in risk that is unacceptable.

Criticality of failure modes can also be assessed by:

- reviewing the severity of the failure effect (on patient, product or process)

- the likelihood that the failure mode might occur and lead the consequence or harm

- detectability of the failure mode

Based on Severity, Occurrence, Detection scores (SXOXD) an RPN, Risk Priority number can be established.

Identify Failure Modes

Failure modes identification:

DFMEA

- o If similar devices exist, failure modes may be known or developed based on existing data and knowledge.
- o Failure modes may present themselves in different modes of operation
- o Failure modes may present themselves at different stages of the product lifecycle
- o Can failure modes be a result of storage or transport
- o Is there material issues that can result in failure modes

PFMEA

- o If a similar or existing manufacturing process is used, the process failure modes may be available
- o Is the process subject to environmental changes or trends
- o Are there failure modes due to operational issues, human error or automation issues

Identify existing controls or detection methods

After failure modes are identified for each process step or element, the existing control measures and any detection methods should be recorded against each failure mode. Controls may prevent a failure mode or reduce its occurrence while detection allows identification of the failure mode which allows reaction or intervention.

Identify effects of failure modes

A failure effect can be understood as a consequence of a failure mode. Failure effects may be caused by one or more failure modes. The description of each failure effect have a level of detail that allows the assessment of the severity level and what the consequences would be.

Identify failure causes

The cause of the failure and how it occurs is helpful in reducing the likelihood of failure or its consequences.

Common cause and common mode failures

Common cause failures happen when more than one element fails at the same time or within a short period of time that result in the effect of the failures. Elements can also fail in the same way or with the same failure mode- however, this can be due to different causes or the same cause.

The likelihood of occurrence can be estimated using:

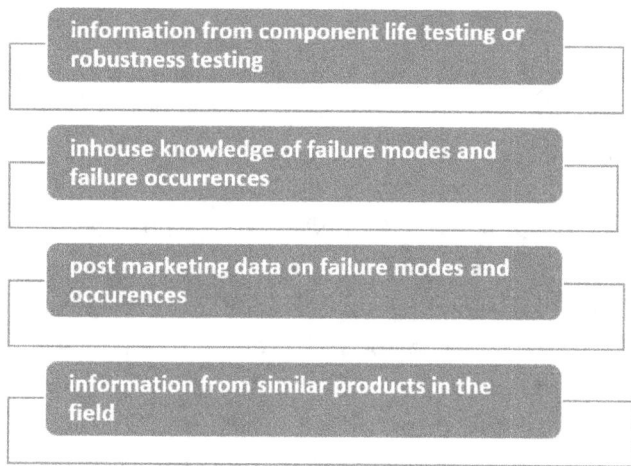

Identify actions

This could include additional controls, revalidated, additional audit requirements of engineering changes.

If new controls or detection methods are agreed as necessary and introduced, re-analysis should be completed post implementation of changes. This is to assess if:

- any new failure modes or effects have been introduced (e.g. re-sterilisation may damage packaging

- the criticality of the particular failure modes is deemed acceptable.

Usability Engineering and Product Design and Development

Recent changes in regulation (e.g. MDR 2017/745) have increased the focus on usability requirements for medical devices. With stronger references to risks associated with use error and foreseeable misuse now requiring manufacturers to respond to foreseeable misuse.

If medical devices are designed and developed without applying a usability engineering or human factors engineering, their use can be non-intuitive, difficult to learn, difficult to complete tasks and difficult to use. Furthermore, as technology and medical devices offer more innovative solutions, patients can now be tasked with using medical devices or administering their treatments, therefore usability becomes more important.

While the goal of design should aim to provide medical devices that are inherently safe, as with most medical devices, residual risks remain once a product is designed, manufactured and validated clinically. Use errors or Usability errors contribute to those potential risk scenarios where medical device usability is an issue for the user. With the increasing abundance of medical devices is the observation, treatment and monitoring of patients, use errors must be assessed for medical devices and reduced to an acceptable level. In contrast to safety inherited by design, the least and often the last protective measures are warnings or contraindications provided on labelling or instructions for use.

The strength of applying usability engineering principles (aka human factors engineering) medical devices is that use errors can be identified and mitigated through design and engineering practices, early-on in the product development and product realization cycle. Medical devices designed and developed devoid of usability engineering, are less intuitive, difficult to use and require focus and attention to learn to use them effectively. In addition, usability is a growing requirement from a regulatory perspective.

The Usability Engineering Process and the subsequent activities should be planned in order to provide a roadmap of deliverables and ensure the requirements are fulfilled. The execution of studies must be executed, documented and approved by appropriate personnel with adequate training, education or experience.

The Usability Engineering Effort for a particular medical device can be estimated based on factors such as:

- Complexity or size of the User Interface, including readability for an IFU
- Complexity of the Use Specification (environment, user etc.)

- Severity of the harm associated with the use of the medical device

In general terms, the following usability questions can be used to understand some of the characteristics of the device.

- Is it easy to learn how to use the device?
- Do users remember how to use the medical device after periods of non-use (days, months)?
- How efficiently can the device be used?
- Is the device designed and manufactured in such a way that prevents users from making errors or allow the user recover from their use errors?
- Is the device appropriate for the user profile, taking into account their abilities?

While safety and performance are the principle concern for medical device manufacturers, usability engineering can be applied looking at non safety related tasks that users may complete. This can be beneficial for the overall user experience and benefit the manufacturer from a commercial point of view.

User needs

Establishing the needs of the user early on in the product development lifecycle can help inform the usability objectives of a medical device. This voice of customer or stakeholder research helps ensure that the product has the right design inputs identified making the development process as effective as possible.

Redesign

During the development of a device, the device or a prototype should be evaluated to determine if user interfaces are correctly specified and designed. Usability testing (formative) is extremely valuable in providing this feedback. If unanticipated use errors are discovered, there is time to address them if it is not close to the design transfer activities. Redesign of design changes resulting from evaluations and observations may be the subject of design change control.

Environmental Factors and Use Error

According to the purpose and intended use of a device, there may be environmental factors that can distract the user or if under challenging circumstances, making it more difficult for the user to not follow the instructions or not operate the device properly. Outdoor settings, hot and cold environments, ambulances or noisy environments are use scenarios that prevent specific challenges.

During usability evaluation, real world settings should be considered in order to obtain the most useful information that can be used as feedback for design teams. Establishing a Use Specification and testing of the User interface are important in this regard.

Quality by Design

Inherent or built in quality that supports safe use of a medical device is the desired approach. When a manufacturer uses the principles of usability engineering and meets the requirements of usability standards, they position themselves for greater success, by identifying usability issues and addressing them by with design solutions. While instructions for use and labelling provide important information and help to mitigate against some risks, the focus should be eliminating or reducing the risks at the device level rather than through documentation.

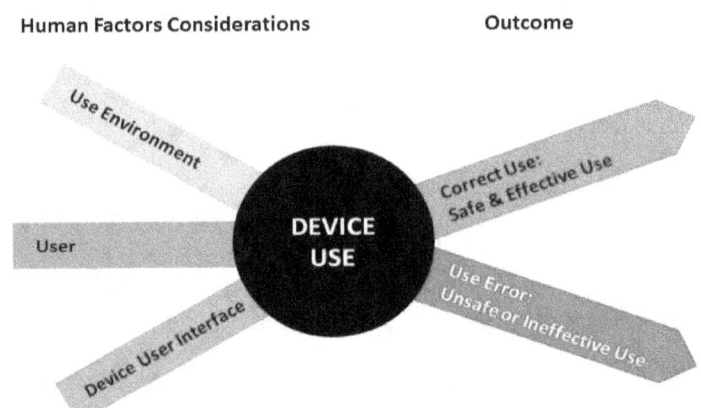

Source: *Guidance for Industry and Food and Drug Administration Staff: Applying Human Factors and Usability Engineering to Medical Devices, FDA.*

Key Terms

Use Error

The formal definition of Use Error is when *"User action or lack of user action (omission) while using the medical device that leads to a different result than that intended by the legal manufacturer or expected by the user"*. It is the "inability of the user to complete a task". It also should be noted that Users may or may not be aware that a use error has occurred.

Use errors can be caused be tasks not been followed per instructions, used in the incorrect environment or not for the intended use or indications. Unexpected physiological responses of the patient is not by itself considered use error. Nor is a malfunction of a medical device that causes an unexpected result. Use error occurs if the user is not able to complete a task. Use errors can result from poor understanding between the principles of operation, the characteristics of the user, user interface, task, or use environment.

A use error occurs at the "action" stage of the interaction with the device. Therefore, at the stage of perception (e.g. misreading a display) or at the stage of cognition (e.g. misinterpreting a number) are not considered use errors. Errors in perception and errors in cognition are classed as contributing factors to or causes of use errors. Hence, a use error (incorrect action or lack of action) can be caused by a misreading or by a misinterpretation of the medical device output, but the use error manifests itself only when an incorrect (aka erroneous action or lack of action takes place.

List of Potential Use Errors

- Error of commission, user performs incorrect action
- Error of omission, incorrectly omitting (failing to complete) a necessary action

List of Potential Factors that can lead to Use Errors

- Environmental distractions
- Excessive workload or stress
- Tiredness
- Lack of attention
- Working fast or too quickly
- Over confidence
- Lack of training
- Lack of experience
- Lack of language abilities (fluency)
- Interruptions

Correct Use

Correct use is when the user successfully uses the medical device and no use errors are encountered.

Normal use

During Normal use, a use error can occur while the user attempts to use the medical device in accordance with its instructions for use. Normal use encompasses all foreseeable user actions when a user is operating a medical device according to the manufacturer's intended use. Normal use is simply what is expected from a user under normal conditions of use, which includes actions that are either correct or in error.

Normal use is differentiated from intended use. Intended use addresses the medical purpose while normal use includes the medical purpose but also the storage, transportation, maintenance (if applicable) and so on.

Abnormal use is defined separately.

Abnormal use
Abnormal use includes use of the device with exceptional disregard for the intended use and instructions or disregarding the contraindications.

Reckless use
Reckless use involves situations where the user is not concerned with the potential danger of their actions.

Sabotage
Normally understood as related to a formative or summative evaluation where post testing the user admits they made a conscious decision to ignore instructions or not to complete an action.

Identification of Use Errors

The process of conducting Usability (Engineering) studies plays a key role in identifying scenarios where reasonably foreseeable misuse occurs.

Use Error is defined as a "user action or lack of user action while using the medical device that leads to a different result than that intended by the manufacturer or expected by the user"

Use Difficulty

Use Difficulties include repeated attempts to complete a task such as:

- ✓ hesitating,
- ✓ excessive "exploring" of the interface
- ✓ unexpectedly referring to the labeling information

Close Call

When a user makes a Use Error but then takes an action to "recover" and prevent the harm from occurring. Close calls may highlight problems with the design of the user interface.

Success

Usability testing or usability engineering studies can be performed during the development of a new product. It acts as a verification that a device is designed appropriately and can identify scenarios or conditions that users could present a use error or usability risk to the patient or user.

As defined above, Use Error is defined as a "user action or lack of user action while using the medical device that leads to a different result than that intended by the manufacturer or expected by the user" Technical report, ISO TR/24971:2020. This covers the following errors:

-the inability of the user to complete a task.

-Use errors resulting from a mismatch between the characteristics of the user, user interface, task, or use environment. Users may be aware or unaware that a use error has occurred.

Exception (to a use error):

- An unexpected physiological response of the patient is not by itself considered use error.

-A malfunction of a medical device

Critical task

A user task which, if performed incorrectly or not performed at all, would or could cause serious harm to the patient or user, where harm is defined to include compromised medical care. A task usually has a specific goal in mind.

Effectiveness

A measure of the accuracy and completeness in which a user achieves specified goal or outcome.

UOUP, User Interface of Unknown Provenance

User interface or a part of user interface previously developed where adequate records of the usability engineering process of recognized standards are not available.

Usability Engineering Process

Each manufacturer should establish a Usability Engineering Process within their organization. This is necessary to embed usability engineering in product development and through the lifecycle of medical device produces. The core purpose of any Usability Engineering Process is to make the medical device user interaction safer. This requires usability problems and use errors to be identified and mitigated by ensuring all known or foreseeable hazard-related use scenarios are addressed.

The strengths of adapting a process approach to engineering is well regarded and evident by the principles of ISO 9001 and ISO 13485. The Usability process must not only be established but needs to be implemented (for each product), maintained and known to be effective during the lifecycle of a product. For a process to properly function, there are some essential activities needed. However, for a process to be effective it depends on its level of integration with other related processes and data generated from processes. For example, the usability process depends on post marketing surveillance data been generated and allowing usability issues to be identified.

Estimating and evaluating the associated risks, controlling those risks, and monitoring how effective those controls is a continual process that occurs throughout the life-cycle of a product.

A usability process assists a manufacturer in the analysis, specification, design and evaluation of the usability of a medical device. The risk management process been interrelated to the usability takes the usability of the device into account. Use errors are minimized by:
    a) discovering hazards and hazardous situations related to the user interface
    b) designing and implementing measures to control the risks related to the user interface
    c) evaluating the risk control measures.

All known or foreseeable hazard-related use scenarios are addressed prior to selecting those hazard-related use scenarios which are then used in preparing the user interface evaluation plan. It is also useful to be mindful of the meaning of a safe medical device. Safety can be understood as freedom from unacceptable risk. Unacceptable risk can arise from use error, which can lead to exposure to direct physical hazards or loss or degradation of clinical functionality.

Formative Evaluation

Formative studies or evaluation is performed early-on and throughout the develop process using simulations and early working prototypes that explore if general usability principles and intended use can be performed. These formative studies can inform and help develop suitable user interfaces and identify is any design changes are required.

Summative Evaluation

Summative Evaluations are completed during the design validation stage of a project. As a validation activity, it should have formal acceptance criteria. Summative testing is typically more detailed and specific than formative testing

Inputs into Risk Identification

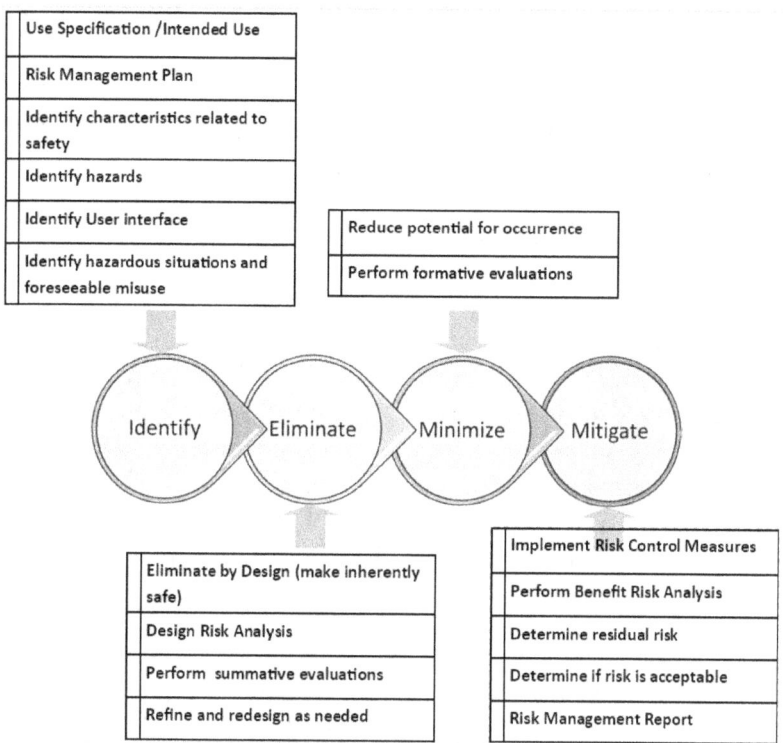

The purpose of any usability engineering process is to identify use errors and eliminate or minimize the use associated risks.

Preliminary Evaluations

Preliminary analyses and evaluations are performed to identify user tasks, user interface components and use issues early in the design process. These analyses help focus the HFE/UE processes on the user interface design as it is being developed so it can be optimized with respect to safe and effective use.

One of the most important outcomes of these analyses is comprehensive identification and categorization of user tasks, leading to a list of critical tasks. Human factors and usability engineering offer a variety of methods for studying the interactions between devices and their users. Analytical methods and empirical methods can be useful for identifying use-related hazards and hazardous situations. These techniques are discussed separately; however, they are interdependent and should be employed in complementary ways. The results of these analyses and evaluations should be used to inform your risk management efforts and development of the protocol for the human factors validation test.

Critical Task Identification and Categorization

An essential goal of the preliminary analysis and evaluation process is to identify critical tasks that users should perform correctly for use of the medical device to be safe and effective. You should categorize the user tasks based on the severity of the potential harm that could result from use errors, as identified in the risk analysis. The purpose is to identify the tasks that, if performed incorrectly or not performed at all, would or could cause serious harm. These are the critical tasks. Risk analysis approaches, such as failure modes effects analysis (FMEA) and fault tree analysis (FTA) can be helpful tools for this purpose. All risks associated with the warnings, cautions and contraindications in the labeling should be included in the risk assessment.

Reasonably foreseeable misuse (including device use by unintended but foreseeable users) should be evaluated to the extent possible, and the labeling should include specific warnings describing that use and the potential consequences. Abnormal use is generally not controllable through application of HFE/UE processes. The list of critical tasks is dynamic and will change as the device design evolves and the preliminary analysis and evaluation process continues. As user interactions with the user interface become better understood, additional critical tasks will likely be identified and be added to the list. The final list of critical tasks is used to structure the human factors validation test to ensure it focuses on the tasks that relate to device use safety and effectiveness.

Identification of Known Use-Related Problems

When developing a new device, it is useful to identify use-related problems (if any) that have occurred with devices that are similar to the one under development with regard to use, the user interface or user interactions. When these types of problems are found, they should be considered during the design of the new device's user interface. These devices might have been made by the same manufacturer or by other manufacturers. Sources of information on use-related problems include customer complaint files, and the knowledge of training and sales staff familiar with use-related problems. Information can also be obtained from previous HFE/UE studies conducted, for example, on earlier versions of the device being developed or on similar existing devices. Other sources of information on known use-related hazards are current device users, journal articles, proceedings of professional meetings, newsletters, and relevant internet sites, such as:

- FDA's Manufacturer and User Facility Device Experience (MAUDE) database
- FDA's MedSun: Medical Product Safety Network
- CDRH Medical Device Recalls
- FDA Safety Communications
- ECRI's Medical Device Safety Reports
- The Institute of Safe Medical Practices (ISMP's) Medication Safety Alert Newsletters

Usability Engineering Process Overview

Prepare Use Specification
- Include intended use and indications
- intended user profile and patient population
- Intendened body part of tissue device interacts with
- Use environment
- Operating Principle

Identify user interace characteristics and potential use errors
- What are the primary operating functions
- How does the user interface address safety and use errors

Identify known of foreseeable harzards and hazardous situations
- Identify hazards that could arise that affect patients
- Use Risk analysis tools to identify use errors

Identify Harzard-related Use scenarios
- List hazardous situations with hazardous Use scenarios
- Record tasks for use scenario and the severity of harms (E.g. Use-related risk assessment)

Create User Interface Specifiation
- Consider content of Use specification, known or foreseeable use errors.
- User interface specifiction should identify testable technical requirements and if training/documentation is required

Create User Interface Evaluation Plan
- Identify the method and scope of formative and summative studies, user profiles, test environment

Perform User Interface Formative Evaluation
- Via usability engineering methods, complete formative evalutions to assess the design. Revise is new user errors or hazards are identified

Perform Summative Evaluation
- Improve User interface if required
- If not practicle provide rationale
- determine residual risk

Usability and Risk Management

A manufacturers approach to Risk Management forms an essential part of regulatory requirements but also the identification of risks and how they are treated. In this respect, when risk management is applied diligently, it makes for a safer medical device where performance issues and potential hazards can be addressed in the design of the device in a preventative manner. Risks and hazards associated with medical devices can be divided into two separate categories.

1) Device failures that are unrelated to usability. For example;

- mechanical
- physical
- chemical
- functional failures
- biochemical
- Sterility

2) Use-Related Hazards by as a result of action on lack of action by the user resulting in a harm.

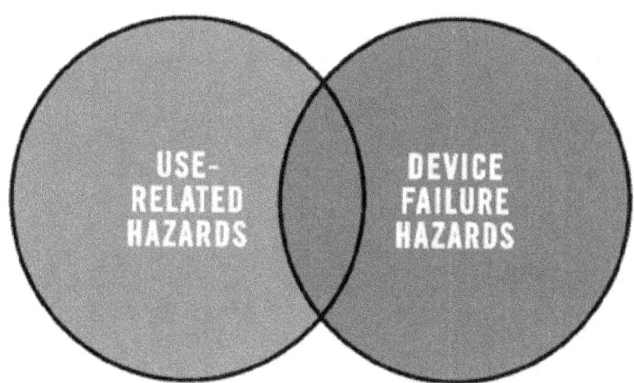

Source: *Guidance for Industry and Food and Drug Administration Staff: Applying Human Factors and Usability Engineering to Medical Devices, FDA.*

User errors are different to component failures of failures in functionality. It is known that estimating the probability of use errors occurring is a challenge, therefore focusing on the severity of the potential have is more important that the combination of severity and probability of the error occurring.

IEC 62366-1 Medical Devices, Application of usability engineering to medical devices is referenced both in ISO 14971 (Medical Devices- Application of Risk Management) and

ISO/TR 24971 (Guidance of the application of risk management). While Risk Management and Usability Engineering are separate processes, they both supplement and overlap in their intent.

Usability engineering can determine if a certain misuse is reasonably foreseeable or not. The completion of a usability test can identify when users may frequently use the medical device in a manner that does not follow the manufacturer's instructions for use. Causes of misuse identified in testing can be a result of several factors such as:

- Poor working culture
- inadequate perception of risk
- limited knowledge of the consequences
- operating procedures /instructions are not clear

Therefore, risk management provides the tools and a decision-making process for estimating and evaluation these usability risks and determining if risk reduction is required or if the potential hazards or hazardous situations can be deemed acceptable risk Foreseeable hazards and risks can be a result of the design, functionality, layout or complexity of the user interface.

- ISO 14971 requires that risks associated with each of the identified hazardous situations be estimated and evaluated
- The manufacturer must establish a risk acceptability policy and criteria

- If a risk is not acceptable using the manufacturer's risk acceptability, appropriate risk control measure(s) that reduce the risk(s) to an acceptable level

- These controls must be implemented and verified as effective in reducing the risk to a pre-determined acceptable level.

Risk Management of Use Errors

This section covers techniques that identify and control use error in devices:

- Define Intended Use, User and Use Environment
- Identify Use Related Hazards
- Analyze Device Use Tasks
- Estimation of Use Related Hazards
- Implement Risk Controls
- Validate Safe Use of Device
- Determine Risk Acceptability

The above elements are described in the following section.

Define Intended Use, User and Use Environment

- The Intended use must be defined for the product early on as this represents the beginning of risk management. The intended use also informs the required design inputs
- The intended use is a short description of the medical device and its purpose
- The user must be defined bearing in mind the use scenarios also. A user may be a lay person, physician, nurse or other healthcare professional. The definition of the user provides perspective on what level or training and technical knowledge the user may have.
- The Use environment may require specific design inputs to protect products from influences of the operational environment. More specifically for usability, depending on the use scenario, the manufacturer and design may need to take into account factors such as stress, noise, outdoor use. This knowledge and context can then inform the user interface design to make interaction more intuitive for the user and minimise use-error.

Identify Use Related Hazards

- Identification of use-related hazards should commence early during the development of a product. The identification of use related hazards can be achieved via task analysis, input from stakeholders and users and formative and summative studies.
- Analysis of similar devices can provide a source of potential use errors and their occurrence levels

Analyze Device Use Tasks

- This technique helps to identify the user tasks in detail. Once critical tasks are identified, corresponding user requirements and user interface requirements can be realized.
- For each user requirement, potential failure modes from a usability perspective.

Estimation of Use Related Hazards

- Estimating the severity and occurrence of use errors can assist in organising and prioritising use-related risks and hazards.
- Failure modes effects and analysis is a risk assessment technique that ranks risks in terms of severity, occurrence and detection. The scores of the severity x occurrence x detection provide a Risk Priority Number, RPN. Other methods of risk assessment include Fault Tree Analysis

Implement Risk Controls

- Identified hazards are preferably eliminated using design features. This may require design modification to a current prototype. These changes should be agreed by a cross functional team. In addition, it should be highlighted that engineering or design features that may control use-errors can themselves present other risks. E.g. mechanical or software failure. However, not all use-errors can be eliminated by design changes and residual risks remain.
- In certain circumstances a change to the intended use or redefining the target or user population can be an option to the manufacturer if other protective measures or mitigations are not feasible.
- Training requirements on the correct use of the medical device should be considered. Training is susceptible to been less effective over time.
- Warning statements, symbols and labelling with they provide a degree of protective measures, their effectiveness depends on whether or not the user notices, reads, understands and follows the meaning of such controls.

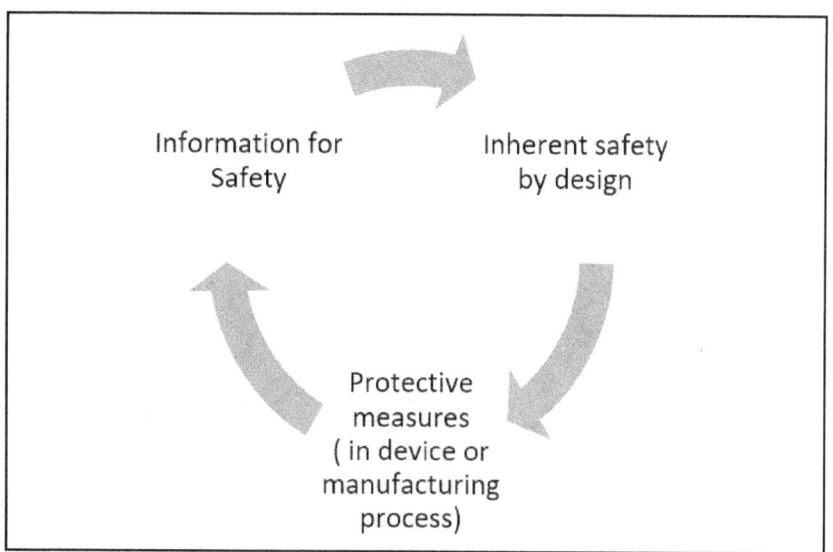

Validate Safe Use of Device

- Device safety can be best demonstrated by validation where the validation takes into account the intended use been applied by the user in a realistic environment.
- The validation should use an appropriate sample of users that has statistical rationale
- The final (or final stage) device should be used in conducting the validation in order to provide reliable results

Determine Risk Acceptability

- After applying a usability engineering process where use specification, user profiles, Use Related Risk analysis; have been completed iteratively through the design cycle, and formative and summative evaluations are concluded and where appropriate, actions taken, the residual risks for use-error should be very low. A determination on whether the remaining risk is acceptable must be made in the wider risk management context and in keeping with the manufacturers risk acceptability policies.

Risk Management Plan

A risk management plan describes the scope of the risk management activities (methods) along with the responsibilities and authorities, verification activities, verification methods, the production and post-production information to be collected and reviewed for the medical device and the criteria for risk acceptability.

Risk Acceptability Matrix						
P5 Frequent	Acceptable	Acceptable	Acceptable	Unacceptable	Unacceptable	Unacceptable
P4 Probable	Acceptable	Acceptable	Acceptable	Unacceptable	Unacceptable	Unacceptable
P3 Occasional	Acceptable	Acceptable	Acceptable	Acceptable	Unacceptable	Unacceptable
P2 Remote	Acceptable	Acceptable	Acceptable	Acceptable	Unacceptable	Unacceptable
P1 Rarely	Acceptable	Acceptable	Acceptable	Acceptable	Acceptable	Unacceptable
	S0 None	S1 Negligible	S2 Minor	S3 Serious	S4 Critical	S5 Catastrophic

The risk management plan must be reviewed and updated throughout the lifecycle of the medical device as new information becomes available. Any changes to the risk management plan must also be recorded in the risk management file. An important factor to consider with regard to the level of detail should be commensurate with the level of risk associated with the medical device.

Identification of hazards from use errors

The usability testing or studies can highlight if issues occur when the device is used by the patient- for example, do people use the medical device in a way that it is not intended to be used or not in accordance with the instructions for use.

Hazards from reasonably foreseeable misuse

Some hazards and hazardous situations may be a result of reasonably foreseeable misuse. Engineering usability studies can also help identify and confirm reasonably foreseeable misuse scenarios.

Identification of hazards relating to Usability

The below series of questions are accompanied with an explanation and factors that should be considered when identifying hazards.

1) **Is successful application of the medical device dependent on the usability of the user interface?**

Depending on the function and intended use of the device most devices shall require certain critical tasks to be done in order to use the device. The usability of the user interface can determine if use-errors are likely to occur. A well-defined, developed and evaluated user interface minimizes greatly the risk of use errors.

2) **Can the user interface design features contribute to use error?**

Factors that should be considered include: control and indicators, symbols used, ergonomic features, physical design and layout, hierarchy of operation, menus for software-driven medical devices, visibility of warnings, audibility of alarms, standardisation of colour coding.

3) **Is the medical device used in an environment where distractions can cause use error?**

   Factors that should be considered include:
   — the consequence of use error;
   — whether the distractions are commonplace;
   — whether the user can be disturbed by an infrequent distraction;
   — whether repetitive stress can reduce the user's awareness or attention.

4) Does the medical device have connecting parts or accessories?

   Factors that should be considered include the possibility of wrong connections, similarity to other products' connections, connection force, feedback on connection integrity, and over- and under tightening.

5) **Does the medical device have a control interface?**

   Factors that should be considered include spacing, coding, grouping, mapping, modes of feedback, blunders, slips, control differentiation, visibility, direction of activation or change, whether the controls are continuous or discrete, and the reversibility of settings or actions.

6) **Does the medical device display information?**

   Factors that should be considered include visibility in various environments, orientation, the visual
   capabilities of the user, populations and perspectives, clarity of the presented information, units, colour coding, and the accessibility of critical information.

7) **Is the medical device controlled by a menu?**

   Factors that should be considered include complexity and number of layers, awareness of state, location of settings, navigation method, number of steps per action, sequence clarity and memorization problems, and importance of control function relative to its accessibility and the impact of deviating from specified operating procedures.

8) **Is the successful use of the medical device dependent on a user's knowledge, skills and abilities?**

   Factors that should be considered include:
   — the (intended) users, their mental and physical abilities, skill and training;
   — the use environment, ergonomic aspects, installation requirements;
   — the personal characteristics of intended users that can affect their ability to successfully interact with the medical device.

9) **Will the medical device be used by persons with specific needs?**

   Factors that should be considered include:
   — users with special characteristics, such as disabled persons, the elderly and children, who might need assistance by another person to enable the use of a medical device;
   — users having wide-ranging skill levels and differing cultural backgrounds and expectations that could lead to differences in what is considered appropriate application of the medical device.

10) **Can the user interface be used to initiate unauthorised actions?**

Factors that should be considered include whether the user interface allows the user to enter an operation mode with restricted access (e.g. for maintenance or special use), which increases the possibility of use error and thereby the associated risks, and whether the user becomes aware of having entered such operation mode.

11) **Does the medical device include an alarm system?**

Factors that should be considered are the risk of false alarms, missing alarms, disconnected alarm systems, unreliable remote alarm systems, and the user's ability of understanding how the alarm system works.

12) **In what ways might the medical device be misused (deliberately or not)?**

Factors that should be considered are incorrect use of connectors, disabling safety features or alarms, neglect of manufacturer's recommended maintenance, unauthorized access to the medical device or to medical device functions.

13) **Is the medical device intended to be mobile or portable?**

Factors that should be considered are the need for grips, handles, wheels or brakes, and the need for mechanical stability and durability.

Use Related Risk Analysis

Use Related Risk Analysis (URRA) is a risk analysis format and methodology that is used to identify and analyze hazards and harms arising from user errors and unintended use of a product/device. The Use Related Risk Analysis (URRA) is part of the risk management process for the medical device and forms part of the risk management file.

The initial use risk assessment should be performed as patient or user needs and inputs are developed. In the event new risks are identified that require mitigation follow up evaluations should be completed to identify protective measures.

After a product is launched and commercialized, the URRA still is available element of risk management. Post-Production Surveillance information collected will be evaluated for Use Errors and the Use Related Risk Analysis is updated as necessary.

IEC 62366-1 Medical devices — Part 1: Application of usability engineering to medical devices

IEC 62366-1 is an international standard that provides a framework to medical device manufacturers to analyse, specify, develop and evaluate the usability of a medical device as it relates to safety in a process-based approach. As described previously, the usability engineering process works to assist the manufacturer in assessing and mitigating risks associated with correct use and use errors during normal use.

IEC 62366 specifies a framework and process for a manufacturer to analyse, specify, develop and evaluate the usability of a medical device as it relates to safety. The systematic evaluation of devices for usability (aka human factors) is intended to identify and minimise use errors and in turn reduce use-associated risks. Once use errors are identified it allows the manufacturer to respond and mitigate as necessary.

The usability engineering (human factors engineering) process as prescribed in IEC 62366 allows a manufacturer to assess and mitigate risks associated with correct use and use errors, i.e., normal use. it can be used to identify but does not assess or mitigate risks associated with abnormal use.

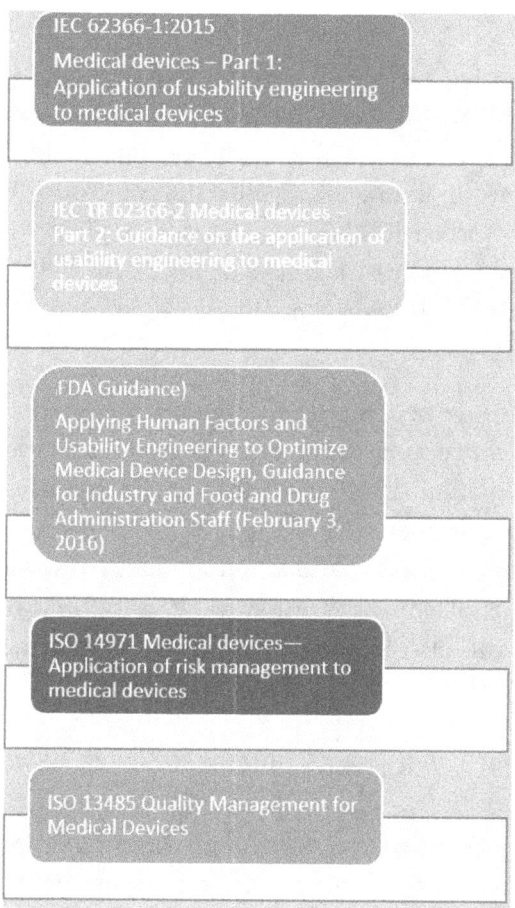

ISO 13485 & Product Realization

Product realization per ISO 13485 is the process of planning product development and introduction but also the subsequent steps that are meaningful in the success of the product introduction. Planning should be initiated early-on in the design stage and should include timelines, resources required, intended markets. One of the most important aspects of planning is gaining the correct stakeholder needs.

Product realization must establish customer requirements and document the design and development efforts. ISO 13485 also has requirements around purchasing, production, service product and monitoring and measuring equipment. Product realization can be defined as a collection of processes that delivers a product or service to the customer. There is an 'opt out' mechanism that where an organization can exclude specific requirements, in cases where product realization is not applicable.

Planning of Product Realization / Design and Development Planning

It is the manufacturer's responsibility to establish and maintain plans that describe or reference the design and development activities and define responsibilities for implementation. The plans should identify and describe the interaction with different groups or activities that are part of the design and development process. The maintenance of plans to reflect an accurate state as the design and development progresses is also a key factor. The design and development planning is intended to be prospective in nature. It allows risks to be identified earlier and promotes timely delivery of projects.

Product realization refers to the product development process that forms part of a product's life cycle, beginning with a conception, through its development, design iterations, verification, validation and to final completion. Product realization for medical devices is defined in ISO 13485 as a mandatory requirement for a quality management system (QMS). Product realization when adopted by a company and when applied consistently can provide benefits to the customer and the manufacturing company. It supports regulatory compliance and successful applications but can also better deliver products that meet customer and company requirements.

7.1 PLANNING OF PRODUCT REALIZATION

Planning is the first step in product realization and provides stakeholder alignment and a roadmap to success. Planning your product realization requirements should include:

- Marketing requirements that may specify product size, features, technology, colour, cost
- Targeted Quality requirements for the product
- Timelines for delivery of project milestones
- Product Intended Use and User Profiles
- Information on product verification and validation
- Applicable standards to be met e.g. Usability, sterility, product specific standards

7.2 CUSTOMER-RELATED PROCESSES

This requirement mandates organizations to fulfill all customer needs, whether those are explicitly stated by clients, necessary for an intended use, or required by regulatory standards. It also covers the best practices for communication of customer feedback. According to the ISO 13485 standard, a customer-related process extends further to delivery, post-delivery activities, and user training requirements.

7.3 DESIGN AND DEVELOPMENT

As designs go through different iterations and modifications, they must continuously reviewed against the inputs, with the outputs maintained as accurate and reflective of the design. It covers the following:

- Intended use of the medical device
- Requirements of regulation, quality and standards
- Product validation and verification
- Design transfer
- Design History requirements
- Control of Design changes

7.4 PURCHASING

The overall performance and quality of the supply chain management must be considered in the context of purchasing, selecting vendors during product realization.

7.5 PRODUCTION AND SERVICE PROVISION

Per ISO 13485, the production process should be planned, executed, monitored, and controlled so that it meets your specifications. Service provision may not be applicable to all products.

7.6 CONTROL OF MONITORING AND MEASURING EQUIPMENT

The control and monitoring of product is supported by measurement and inspection equipment. Ensuring this equipment is controlled, monitored and maintained to be reliable and consist is important in delivering safe and effective products.

Activities such as calibration should be completed in accordance with documented procedures, external standards and where nonconformances occur, a process for controlling product and addressing issues is available.

Usability Engineering Plan

A Usability Engineering Plan (UEP) should be established during the planning stage of product design and development. Sometimes the plan is stand alone or alternatively it can be included in the risk management plan.

While FDA regulation, 21 CFR Part 820 does not require a Usability Engineering Plan/ Human Factors plan, it is useful to outline the overall usability effort, the deliverables and where the activities fall within the overall design and development plan.

Content of Plan:

- Introduction- outlining the product, purpose and scope of the plan
- Roles and Responsibilities to ensure the right stakeholders review and approve
- User Research- user needs assessment, literature review, competitor analysis, task analysis
- User Interface Design- the key characteristics required, critical to quality and other design and usability requirements of the medical device
- Risk Management- Description of how risks are identified and how risks to users is reduced via Usability engineering studies
- Usability Testing- outline of formative and summative evaluations planned

Product Realization Process and Risk Management

Manufacturing companies that are responsible the production, design and development of medical devices are required to have processes and procedures in regards not only to risk management but also Product realization. Risk management and Process realization are normally separate processes with different procedures and SOPs. However, regulations e.g. EU MDR, require that the two processes work together with design and development taking into account risk management. Above all, this is to ensure safety requirements are included in development process and that risk are identified and tracked during the development lifecycle in order to ensure they are addressed or mitigated. The review of the results of the design verification activities during development to verify the risk controls were effective is part of this process.

Formative Studies and Evaluation

Formative studies or evaluation is performed early-on and throughout the develop process using simulations and early working prototypes that explore if general usability principles and intended use can be performed. For medical devices, the main focus of formative studies is on providing a preliminary analyses of the user interface to identify the use safety aspects of user interaction with the device.

Therefore, by completing formative studies early in the product life-cycle, they can inform and help develop suitable user interfaces and identify if any design changes are required. It also serves to uncover any unanticipated use errors. Completing user interface testing via formative evaluations improves the likelihood that the final (summative) evaluation of the usability of the user interface can be completed successfully. Any design changes or modifications to the user interface required should be subject to design control and principles of change management. This is to ensure that changes are reviewed and approved by the appropriate functions and that continuity is maintained. Further evaluations are not required if the specified quality level has been achieved that gives the confidence that the final acceptance criteria will be met when the summative evaluation is conducted at design transfer.

Formative evaluations can include, but are not limited to:

- Hands-on evaluations
- Cognitive Walk-Through
- Customer Surveys
- Focus Groups

The effectiveness of the user interface is challenged by testing the device via formative evaluation. This formative testing can confirm that the user needs and design inputs are fulfilled. It also can identify unknown or unanticipated use errors. The testing done under evaluation may can also assess the effectiveness of risk control measures.

Once a Use error is identified, a root cause should be determined to establish if the use error was a result of an unidentified user need or a weakness in the design of the user interface.

Generally, success depends on completing an evaluation of the user interface in a study with pre-determined acceptance criteria according to the user interface specification. Residual risks related to usability must be controlled to acceptable levels. The manufacturer can apply the acceptance criteria in accordance with their risk policy and ISO 14971.

Design Validation can be demonstrated from a Summative Evaluation report on Human Factors Engineering/ Usability Validation

Formative evaluations can involve simple mock-up devices, preliminary prototypes or more advanced prototypes as the design evolves. They can also be tailored to focus on specific accessories or elements of the user interface or on certain aspects of the use environment or specific sub-groups of users. Design modifications should be implemented and then evaluated for adequacy during this phase of device development in an iterative fashion until the device is ready for human factors validation testing. User interface design flaws identified during formative evaluation can be addressed more easily and less expensively than they could be later in the design process, especially following discovery of design flaws during human factors validation testing. If no formative evaluation is conducted and design flaws are found in the human factors validation testing, then that test essentially becomes a formative evaluation. The effectiveness of formative evaluation for providing better understanding of use issues (and preventing a human factors validation test from becoming a formative evaluation) will depend on the quality of the formative evaluation.

Depending on the rigor of the test you conduct, you might underestimate the existence or importance of problems found, for example, because the test participants were unrealistically well trained, capable, or careful during the test.

Unlike human factors validation testing, company employees can serve as participants in formative evaluation; however, their performance and opinions could be misleading or incomplete if they are not representative of the intended users, are familiar with the device or are hesitant to express their honest opinions. The protocol for a formative evaluation typically specifies the following:

- Evaluation purpose, goals and priorities
- Portion of the user interface to be assessed
- Use scenarios and tasks involved; • Evaluation participants
- Data collection method or methods (e.g., cognitive walk-through, observation, discussion, interview)

A simple kind of formative evaluation involving users is the cognitive walk-through. In a cognitive walk-through, test participants are guided through the process of using a device. During the walk-through, participants are questioned and encouraged to discuss their thought processes (sometimes called "think aloud") and explain any difficulties or concerns they have.

Simulated-Use Testing Simulated-use testing provides a powerful method to study users interacting with the device user interface and performing actual tasks. This kind of testing involves systematic collection of data from test participants using a device, device component or system in realistic use scenarios but under simulated conditions of use (e.g., with the device not powered or used on a manikin rather than an actual patient). In contrast to a cognitive walkthrough, simulated-use testing allows participants to use the device more independently and naturally.

Simulated use testing can explore user interaction with the device overall or it can investigate specific human factors considerations identified in the preliminary analyses, such as infrequent or particularly difficult tasks or use scenarios, challenging conditions of use, use by specific user populations, or adequacy of the proposed training.

During formative evaluation, the simulated-use testing methods can be tailored to suit your needs for collecting preliminary data. Data can be obtained by observing participants interacting with the device and interviewing them. Automated data capture can also be used if interactions of interest are subtle, complex, or occur rapidly, making them difficult to observe. The participants can be asked questions or encouraged to "think aloud" while they use the device. They should be interviewed after using the device to obtain their perspectives on device use, particularly related to any use problems that occurred, such as obvious use error. The observation data collection can also include any instances of observed hesitation or apparent confusion, can pause to discuss problems when they arise, or include other data collection methods that might be helpful to inform the design of a specific device user interface.

In summary, formative evaluation can reveal previously unrecognized use-related hazards and use errors and help identify new critical tasks. It can also be used to:

- Inform the design of the device user interface (including possible design tradeoffs)
- Assess the effectiveness of measures implemented to reduce or eliminate use-related hazards or potential use errors
- Determine training requirements and inform the design of the labeling and training materials (which should be finalized prior to human factors validation testing)
- Inform the content and structure of the human factors validation testing.

The methods used for formative evaluation should be chosen based on the need for additional understanding and clarification of user interactions with the device user interface. Formative evaluation can be conducted with varying degrees of formality and sample sizes, depending on how much information is needed to inform device design, the complexity of the device and its use, the variability of the user population, or specific conditions of use (e.g., worst-case conditions). Formative evaluations are used to inform device user interface design while it is in development

Efficiency

The concept of efficiency in relation to usability can be understood to be the effectiveness in relation to resources expended. The greater the efficiency with a user often leads to better outcomes and safe use. A lack of efficiency can contribute to risks or increase existing risks. If the medical device has a time based or time related performance characteristic, efficiency may also be of greater importance. An obvious example of a medical device where efficiency of its use is critical to outcomes is Automated External Defibrillators (AEDs).

Usability at Design Intent

Usability is created by characteristics of the user interface that facilitate use, i.e. to make it easier for the user to perceive information presented by the user interface, to understand and to make decisions based on that information, and to interact with the medical device to achieve specified goals in the intended use environments. many of these factors can influence safety to various extents.

Safety

Safety is freedom from unacceptable (use-related) risk.

Whereas, freedom from discomfort device is called 'satisfaction'. The manufacturer must distinguish between safety risks and customer feedback relating to satisfaction of positive experience

Usability Engineering File

The usability engineering file is the collection of documents and records that a manufacturer generated in relation to usability and shows the results of the Usability Engineering Process. The Usability Engineering File allows more efficient auditing.

The Usability Engineering File typically includes:
1) Usability Plan (or integrated into the risk management plan)
2) Use Specification
3) UOUP assessment as applicable
4) Use Related Risk Assessment (URRA)

5) User Interface Specification (can be part of the IOV)
6) Usability Report (can be part of the risk management report)
7) UOUP assessment as applicable
8) Formative Evaluations
9) Usability Engineering Report
10) Post Marketing Surveillance

Updates, as appropriate, should be considered throughout the Usability Engineering Process and monitored through the product lifecycle.

Use Scenario

Use scenarios describe the user interaction with the medical device that aims to achieve a certain result under specific conditions of use. Use scenarios can be written as statements, in story like text or by means of simple bullet points that mimic steps or tasks. Different situations can present different use scenarios that can include correct use or normal use with use error in various use scenarios. A hazardous situation is any circumstance in which people, property or the environment is/are exposed to one or more hazards. When a particular use scenario leads to a hazardous situation, the use scenario is called a hazard-related use scenario. An example of a Use scenario would be going swimming while still wearing a 24hr Blood Pressure monitor.

Environmental Factors and Usability

The use environment of a medical device should be considered in the design planning stage of a product and should result in design inputs. A home setting is distinct from a hospital setting. More specifically, various environments co-exist in hospitals but exhibit different environmental conditions that can result in poor use or use errors. Temperature, humidity, noise, limited lighting, confined spaces can all influence the manner the users ability to perform critical tasks.

Use Specification

The use specification is a document that provides a summary of the important characteristics related to the context of use of the medical device. The intended medical indication, patient population, part of the body or type of tissue interacted with, user profile, use environment, and operating principle are typical elements of the use specification.

- descriptions of intended device users, uses, use environments, and training

- intended user population(s) and meaningful differences in capabilities between multiple user populations that could affect user interactions with the device
- intended use and operational contexts of use
- use environments and conditions that could affect user interactions with the device
- training intended for users
- documentation required for users

Examples of Intended Users:
- *Laypersons (patients, lay caregivers)*
- *Nurses*
- *Pharmacists*
- *Doctors*

The Intended patient population must also be considered when developing a Use Specification. This may include the intended age group, health, or condition of the patient population. The intended patient population can be similarly referred to as the user profile which also describes: While the above factors (age, health, condition) are the most important attributes, other information such as Occupation, education level, Linguistic and cultural background and potential disabilities can be considered.

Use Environment

Actual conditions and setting in which users interact with the medical device. Factors that should be considered in describing the use environment include, but are not limited to:
- *Lighting*
- *Sound*
- *Sterile or non-sterile, single-use or reusable*
- *Hospital use or home use*
- *Ward or Operating theatre*
- *Ambulance use,*
- *Transportation*
- *Storage*
- *Disposal*
- *Available personal protective equipment*

User Interface

The user interface encompasses the means of interaction between the medical device (including all of the elements) and the user, either by means of software or hardware interfaces. Documentation such as the instructions for use are considered part of the medical device and its user interface. The terms usability engineering, human factors engineering or 'human factors engineering' can be used. However, applying knowledge of people to the user interface design is better described as human factors engineering. While the activity of evaluating devices and their interfaces may better describe usability engineering.

- *Documentation provided*
- *cables*
- *tubing connections*
- *handles*
- *force required to move the weight*
- *work surface height*
- *markings (labelling)*
- *video display (size)*
- *push buttons*
- *touch screens*
- *auditory signals*
- *vibratory signals*
- *visual signals*
- *keyboard and mouse*
- *haptic controls (knobs, joysticks)*

Guidance on User Interface	
Description of User Interface Specification	The medical device be described in straight forward language and account of any associated connections, parts or assemblies that is required during the operation of the medical device by the user
Example	*For an Upper arm Blood pressure monitor, the user interface consists of the unit which houses the batteries, pump, measurement device and digital display. An upper arm cuff connected to a tube and easy-fit connection*
Factors to consider when	*Use Specification* *All Known or foreseeable use errors associated with the medical device*

developing a User Interface Specification	*Hazard-related use scenarios*
Identify User Needs	*Identify User Needs* *The User Needs are written in terms of device attributes and not user tasks. The intent of the user needs section is to help identify points of user interaction with the device. Should a change to the design be made, the impact on the user interaction will be better understood.* *1. Does the medical device design allow easy handling* *2. If the device is sealed within a sterile barrier, can it be opened aseptically by the user.* *3. Does the packaging provide protection during storage and transportation* *4. If applicable, sterility of device is maintained during shipping, storage and over course of shelf life.* *5. Information printed on the packaging must be locatable and legible by a user with normal eyesight* *6..The device can be removed and assembled easily from the packaging*

User Interfaces of Unknown Provenance (UOUP)

Prior February 2015 products released to the market without any User Interface design changes, are known as User Interfaces of Unknown Provenance (UOUP) per the usability standard- IEC 62366-1 Medical Devices, Application of usability engineering to medical devices. In place of Usability studies and user interface evaluation, a review of Post-Production data can be completed. However, subsequent design changes that impact the usability interface post February 2015 should be evaluated.

User Interface Specification

The user interface specification is made up of the design requirements for a the medical device that document and detail the characteristics of the user interface. User Interface requirements should be **S-M-A-R-T**. Specific, measurable, achievable, relevant and time bound.

User Needs and Requirements

The User Needs can be generated from various sources including focus groups, existing products, and complaint data.

From the User Needs, requirements are generated to meet these needs which have measurable (testable) acceptance criteria.

These requirements can include documentation that is required for safe use of the product which can include Instructions for Use or a User Manual, as well as training that is required. If training is needed who will conduct the training and what training material will be used is to be defined.

Requirement No.	User Interface Requirement
1	Digital display shall be visible at a distance of 1m to three people standing side-by-side, with all able to read the text on the display screen
2	The medical device shall be capable of producing an auditory alarm of 45dba when measured at 1m from the front of the display screen
3	The medical device shall be mobile and have a wheel locking system for when the unit is stationary
4	The medical device shall be compatible with an 230V electrical supply with integrated surge protection via a fused mechanism
5	The operation of the HMI shall support use of a mouse or keyboard to select commands.

The user interface specification is a source of design inputs for the product and should be subject to design control during the development stages.

The User Interface Specification is part of the development of Design Inputs for the medical device and can be captured in the Design (Inputs/Outputs Verification and Validation Matrix (DIOV).

User Interface Evaluation Plan

The User Interface evaluation plan shall identify the objectives and methods of the required formative and summative evaluations. The plan can serve to detail the user interface evaluation activities and other development activities. For example, the Summative Evaluation is completed using product that is representative of the commercial product and therefore, they must be available when the evaluation is executed. In addition, labelling must be representative in order to create simulated conditions.

Summative Testing

Summative Evaluation is used to confirm the safety of the User Interface while also assessing the effectiveness of risk management measures such as risk controls and mitigations. Summative testing form part of the design validation activities in the development stages of a medical device. By definition, and in keeping with the design control process, design validation is usually in the later stages of development.

In comparison to formative testing, summative testing should apply pre-defined acceptance criteria and account for statistical sample sizes and

Summative Evaluation Protocol

- The summative evaluation demonstrates that the intended users of a medical device can safely and effectively perform critical tasks for the intended uses in the expected use environments.
- All aspects of intended use shall be evaluated.
- For the Summative Evaluation the following is to be used: a production version of the device, representative device users, and actual use or simulated use environment.
- The evaluation can be carried out under conditions of simulated use, but, if possible and necessary, it can be undertaken under conditions of actual use in a clinical study.
- If a clinical study is used the conditions must be the same as an actual use including all instructions for use and training.
- Summative evaluation of usability has formal acceptance criteria.
- Documenting the criteria for determining whether the user has successfully completed the tasks associated with the hazard-related use scenarios is required.
- One possible way to express these criteria is that no use error that leads to a use related harm. Another way is that no use errors lead to unacceptable risk of harm.

Tasks and Use Scenarios

The test protocol should describe the user tasks and/or use scenarios containing tasks to be included in the test, information regarding task criticality or relative priority, and the process by which task inclusion and priority were determined.

The test protocol should also provide a rationale for the extent of device use and the number of times that participants will attempt to use the device.

The Test Participants must represent the population of intended users. If the device has more than one population of users, then the validation testing should be designed to evaluate each distinct user population.

For devices intended to be used by more than one group of users that have distinct abilities or use roles, the number of participants is to be determined based on content and complexity of the device being evaluated. It is recommended that at least 15* participants from each user group should participate in the validation testing.

*15 is the number referenced in FDA's Guidance on Applying Human Factors and Usability Engineering to Medical Devices. The training provided to test participants should represent the training that actual users will receive. Retention of training decays over time, therefore, prior to testing a period should elapse following training. The FDA recommends this should not be less than one hour but should represent the actual decay period between training and first use.

The summative evaluation report documents the results of the Summative Evaluation Plan and Protocol.

The root causes of problems identified during validation testing should be evaluated from the perspective of the test participants involved and direct performance data will support this determination. Data analysis should include subjective feedback from participants regarding critical task experience, difficulties, "close calls," and any task failures by test participants.

Failures and difficulties associated with greater than minimal risk and attributable to the user interface should be addressed by designing and implementing risk mitigation strategies and re-testing those elements to confirm their success at reducing risks to acceptable levels without introducing any new risks. Depending on the level of mitigation strategies required, revalidation may be necessary.

<u>Residual Risk</u>

Residual risk is risk that remains that cannot be eliminated or mitigated through any modifications to the product via design, user interface, accessories, labeling or training.

The analysis of residual risk should determine if the residual risk is outweighed by the advantages offered by the device. If design flaws that could harm the patient or user are identified, planning to address them in subsequent versions of the device post launch is not acceptable.

Post-Marketing Surveillance

Post-Production Surveillance (Post-Market Surveillance) consists of customer feedback and complaint data. Customer feedback may simply be due to dissatisfaction while using the device. More serious complaints that have a patient safety impact often require reporting to regulatory authorities. These are known as reportable events or adverse events depending on the region or competent authority. In addition, for a manufacturer to be compliant ISO 14971, post market surveillance data must be reviewed throughout the lifecycle of the medical device.

In support of post-production activities, the manufacturer must establish, document and maintain a system to actively collect and review information relevant to the medical device in the both the production and post-production phases. The methods of data collection and processing should also be included in the product risk management plan.

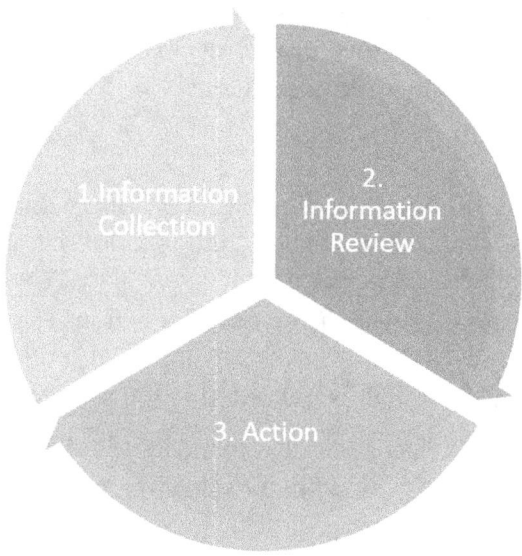

Information collection

The manufacturer must collect, where applicable:

- information generated during production
- information generated by the user (complaints
- information generated by the supply chain;
- publicly available information
- information related to the generally acknowledged state of the art.

Information review

The manufacturer shall review the information collected for possible relevance to safety, especially whether:

- previously unrecognised hazards or hazardous situations are present
- an estimated risk arising from a hazardous situation is no longer acceptable
- the overall residual risk is no longer acceptable in relation to the benefits of the intended use
- the generally acknowledged state of the art has changed. The results of the review shall be recorded in the risk management file

Actions

If the collected information is determined to be relevant to safety, the following actions apply.

Reassessment of risks

- the manufacturer shall review the risk management file and determine if reassessment of risks and/or assessment of new risks is necessary
- Are New Risks or emerging risks identified?

Residual risk

- if a residual risk is no longer acceptable, the impact on previously implemented risk control measures shall be evaluated and should be considered as an input for modification of the medical device;
- the manufacturer should consider the need for actions regarding medical devices on the market
- any decisions and actions shall be recorded in the risk management file.

The collection of data post-production is also mandated by IEC 62366-1. Although all known Use Errors must be identified in the Usability Evaluation, unanticipated Use Errors may not be identified. The information collected post launch can therefore identify any use errors that were missed.

European Regulations- Usability and MDR

GSPR CHAPTER 1, Section 5

In eliminating or reducing risks related to use error, the manufacturer shall:

(a) reduce as far as possible the risks related to the ergonomic features of the device and the environment in which the device is intended to be used (design for patient safety), and

There are several key words and corresponding requirements in section 5 (a). Firstly, there is a requirement to 'reduce as far as possible'. Historically, the term 'ALARP', as low as reasonably possible' was often included in risk management. This is no longer appropriate, and the risk measure must be as low as possible (reasonably has been dropped).

Ergonomic features of the device can be understood to the user interface and human factors engineering. Use scenarios documented by the manufacturer must also account for the environment in which the device is intended to be used.

(b) give consideration to the technical knowledge, experience, education, training and use environment, where applicable, and the medical and physical conditions of intended users (design for lay, professional, disabled or other users).

Part (b) is also intended to eliminate or reduce the risk of use errors by designing products, labelling and instructions for use appropriate user in mind such as a lay person, physician or other.

Per EU MDR 745/2017 a definition of device deficiency includes the term 'user errors' due to an inadequacy.

Article 2, Definitions

(59) 'device deficiency' means any inadequacy in the identity, quality, durability, reliability, safety or performance of an investigational device, including malfunction, <u>use errors</u> or inadequacy in information supplied by the manufacturer;

It is not feasible for a manufacturer to wait until a product is launched to identify use errors. The manufacturer must apply usability engineering to develop use specifications, user interfaces and performance formative and summative evaluations as required. Therefore, the Usability engineering process should identify foreseeable misuses and use errors and via redesign or introduction of protective measures reduce risks to as low as possible, where the benefit outweighs the risks and the risk is acceptable.

Design Controls and Usability Engineering

The FDA Quality system regulation, 21 CFR 820.3 provide manufacturers with Design controls applicable to medical devices. Usability Engineering (Human Factors Engineering, HFE) play an important role in the product development process and the safe design of products.

Design and Development Planning

At the planning stage of a project, the user needs should start to be documented in order to create design inputs. Depending on the severity of harm a device can cause can inform the scope and time of the usability assessment. Therefore, the most important factors to consider for Usability engineering at the D&D Planning stage are user needs (Use specification, User interface), identifying harms and gaining a sense of the overall effort.

Design Input

Usability engineering/ human factors evaluations should be used to define the inputs that relate to user needs. This may be the needs of a patient and/or the user of the medical device.

Design inputs cover not only user needs but factors such as sterility requirements (if applicable), requirements of standards, requirements of local regulation and design for manufacturability. Design controls are usually followed after the first deliverable that will be used to demonstrate compliance to 21 CFR Part 830.30 such as a design and development plan or a requirements document such as a Design inputs outputs verification matrix, (DIOV).

Use-related risk control measures are also design inputs and should be documented through the development of the product

Design Output

Design outputs must be documented and are typically approved by the design teams and stakeholders. Design outputs normally describe an element of the design and provides requirements for the implementation. For example, the physical attributes of a product would be normally detailed in engineering drawings containing measurement values and tolerances. The Engineering drawing is a design output. Design outputs relate back to the design inputs and form evidence that the design input is progressing through the product development process.

Design outputs relating to Usability Engineering include Use-related Risk Analysis' and User Interface specifications.

Design Review

Design reviews are required during the product development process and should occur on a routine basis. Design reviews should include the core team involved in the development of the product and stakeholders. The basis of each design review should include a review of activities against the design and development plan, review of open actions from the previous review and ensure that new requirements or information is assessed and appropriately tracked and added to the project plan.

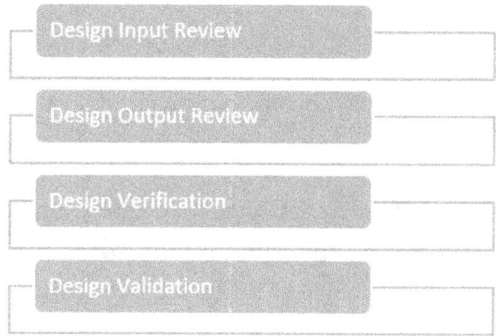

Design Input Review:

- Determine if the design inputs adequately address the user needs and intended use of the medical device?
- Status of the Usability Engineering Plan and are any revisions required?
- Review the actions to be completed in advance of the next design review; e.g. formative evaluation, URRA update.
- Develop user requirements testable for summative testing
- Include Design outputs from formative evaluations in the DIOV.

Design Output Review

- Review results of formative evaluations for use-errors
- Input use errors into risk assessments

- Review any design changes that may impact design outputs
- Review the actions to be completed in advance of the next design review; e.g. formative evaluation, URRA update

Design Verification Review

- Confirm completion of Design verifications
- Verify use related risk controls are implemented or tracked

Design Validation Review

- Summative Evaluations have been completed
- Usability related Risk Control measures are effective
- The benefits outweigh the risks and residual risk is acceptable

Design verification include stability studies, process validations and other studies that confirm design inputs are successfully achieved according to the design output

Design validation involves the completion of studies such as usability evaluations and clinical studies that demonstrate the safe and effective use of a device. The conditions of testing should represent actual or simulated conditions that represent realistic use. The participants of the study should include users within the target population. At this point the development stage, the remaining risk of use-errors should be low, and if present still have appropriate risk controls that are effective.

Design transfer occurs when the device design is required to be translated into a validation manufacturing process for the purposes of commercial manufacturing. To facilitate the production process, design specification, manufacturing specification and finished product specifications are required to be approved and validated to enable commercial manufacturing. At this late stage in the process the Usability Engineering /Human Factors effort should be largely completed with evaluations reported and actions and risk controls validated and effective.

Design changes result in changes to design inputs or design outputs. Design changes include modification to product specifications, material changes, changes to product requirements, changes to product features or functionality. Changes to manufacturing processes can also impact the finished product and as such should be assessed for the impact on the design and functionality of the product.

Therefore, any changes that may impact the usability of a medical device should be treated as a design change and must follow a process of impact assessment/evaluation of the design change, design reverification and revalidation and risk re-evaluation. Evaluation of the design change should ask:

- Does the proposed change impact users?
- Is there an impact on safety?
- Is there an impact on effectiveness?
- Is there an impact on Usability?
- Is there an impact on design verifications or validations?

From a usability engineering perspective, the evaluations form part of the verifications and validations associated with the device design. Therefore, relevant studies should be included in the design history file.

Appendix- 4- Simple format of a Use Related Risk Analysis

Use Related Risk Analysis- Upper Arm Blood Pressure Monitor, Rev 1.0					
User Task	Identify Potential Use Error	Hazard, Harm	Severity	Risk Control Measures	Risk Control Effectiveness Y/N
1. Connecting the air house	Plug on air house not firmly pushed into position	BP reading not possible, ERR_CUFF	1, Inconvenience	Inherent by Device Design- Plug is tapered to provide easy compliance Labelling/Safety information- Instructions provide a labelled diagram IFUXX00X: Precautions and Warnings	Y
		Cuff size	4, lead to	Inherent by Device Design- Arm circumference design meets 80% of patient population	

		inaccurate readings	nsion and/or hypotension	Labelling/Safety information- Arm size is printed on each cuff Instructions provide step by step guidance and diagram of proper fit range IFUXX00X: Precautions and Warnings	
3. Applying the arm cuff	Does not remove clothing covering arm Does not position and orientate the cuff correctly	Prevents accurate readings	4, lead to undiagnosed hypertension and/or hypotension	Inherent by Device Design- Diagram is printed on the cuff Labelling/Safety information- instructions on application IFUXX00X: Precautions and Warnings	Y
4. Start measurement	Removes cuff during measurement Does not maintain position	Prevents accurate readings	4, lead to undiagnosed hypertension and/or hypotension	Inherent by Device Design-inflation of cuff completes in 10seconds Labelling/Safety information- Position and behaviour during measurement detailed in the instructions IFUXX00X: Precautions and Warnings	Y

| 5. Cuff removal | Starts measurement cycle inadvertently | Compression of arm | 2, low to moderate discomfort | Inherent by Device Design-Cuff flap can be removed easily Labelling/Safety information- guidance on remeasurement specified in the instructions for use IFUXX00X: Precautions and Warnings | Y |

Severity Scoring and Descriptions

Severity Score	Description
5	Patient death
4	Permanent harm, if condition left unresolved
3	Moderate injury, no lasting effects
2	Moderate discomfort, no lasting effects
1	Discomfort, transient
0	No harm to user or patient

Information Supplied and Usability

Information for safety is a risk control measure that should be used only after the manufacturer has determined that (further) risk reduction by other measures is not practicable.

Information provided that relates to safety must be assessed through the usability engineering process in order to determine the information is:

- perceivable by users in use environments
- is understandable by users
- allows correct use of the medical device

Risk reduction achieved using modified design features that make the medical device inherently safe is always the preferred course of action. If this is not possible, implementing protective measures can be applied.

Information for safety is instructive and gives the user clear instructions of what actions to take or to avoid, in order to prevent a hazardous situation or harm from occurring. This information can be provided in the form of:
- warnings
- precautions
- contra-indications
- instructions for use
- training

The purpose of Warnings and precautions is to identify adverse reactions and potential safety hazards tat are serious or clinically significant. For adverse reactions to be included in patient information there should be causal association with the product and the adverse event.

A serious adverse reaction resulting in one of the following and should be considered for inclusion in the Warnings and Precautions section of patient literature.
- Death
- Life altering adverse event- significant incapacity
- Hospitalization
- Congenital anomaly or birth defect

Contra-indications should be indicated where the clinical situations results in a risk that outweighs the benefits of the device.

Instructions for use

When developing information for safety, it is important to identify to whom this information is to be provided and how it is to be provided. This can include an explanation of the risk, the consequences of exposure and what should be done or avoided to prevent any harm. The manufacturer should consider:

- regulatory requirement
- the need to classify the information for safety, based on the level of risk;
- the level of detail necessary to convey the information for safety;
- the location for the information for safety (e.g. a warning label on the medical device);
- the wording, pictures or symbols to be used to ensure clarity and understandability;
- the intended recipients (e.g. users, service personnel, installers, patients); the appropriate media for providing the information, (e.g. instructions for use, labels, warnings in
- the user interface);

Information for safety can be communicated in different ways, depending on when in the medical device life cycle the information is to be communicated, e.g. via the user interface of a menu-driven medical device, as cautionary statements in the accompanying documentation, or in an advisory notice.

Information for safety can be given in various forms, such as warning labels attached to the medical device, warning statements in the instructions for use, instructions on a graphical user interface, or instructions in training videos.

- Warning: Do not step on surface
- Warning do not freeze
- Warning Do not refrigerate
- Warning: Do not use if seal is broken
- Warning: Do not remove cover, risk of electric shock

Label Design and User Interaction

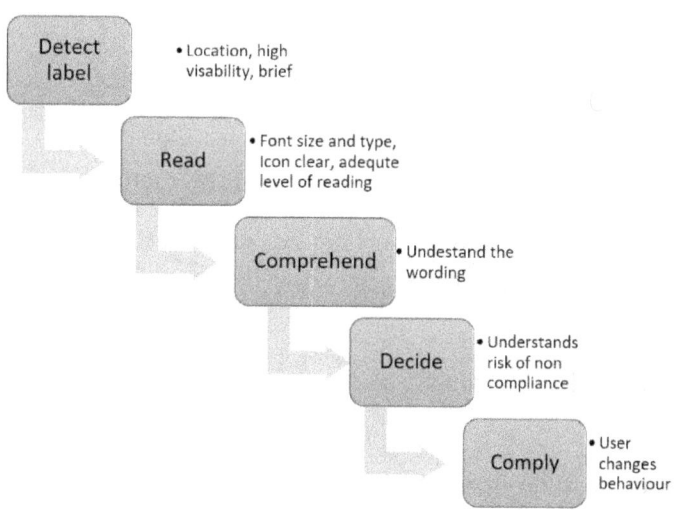

Useful Definitions

Design Verification- Design verification shall confirm that the design output meets the design input requirements.

User- person interacting with (i.e. operating or handling) the medical device.

Layperson-Someone who is not an expert in or does not have a detailed knowledge of a particular subject, in this case the operation or use of the medical device. E.g. the patient or caregiver.

Design Input-a physical and performance requirement of a device that is used as the basis of design.

Design Output-from the results of the design effort and activity, a design output details the device, packaging, manufacturing, testing and labelling requirements

Design Validation- Design Validation is establishing by objective evidence that device specifications conform with user needs and intended use(s).

Input Output Verification Validation Matrix- The traceability matrix created during product development or as part of a design change that defines user needs, inputs, outputs, design verification and design validation.

User Profile-Summary of the mental, physical and demographic traits of an intended user group, as well as any special characteristics, such as occupational skills, job requirements and working conditions, which can have a bearing on design decisions.

User Interface of Unknown Provenance (UOUP)- if the User Interface of a medical device was previously developed (prior to publication of standards) for which adequate records of usability process and Summative Evaluation is not available.

APPENDIX - Risk Questionnaire

Questions 1-7

What is the intended use and how is the medical device to be used?

The intended use (purpose) of a device is critical to identify the safety and performance requirements. From a risk perspective, the intended use assists in identifying potential safety risks, hazards and hazardous situations that may occur during the use or operation of the device. The intended use must be clear and unambiguous. The intended use provides a description of the indications and general application and guidelines for use. It should be well defined and be reviewed by regulatory and medical professionals to ensure compliance with legislation and clinical relevance.

> European legislation for medical devices define 'performance' means the ability of a device to achieve its intended purpose as stated by the manufacturer, Ref. 2017/745 EU MDR.

Is the medical device intended to be implanted?

Depending on whether a device is implantable or not impacts upon the potential risks and complexity of design considerations and performance requirements. However, both implantable and non-implantable devices are subject to risk management and the process of risk identification and so forth is applicable to both.

Is the medical device intended to be in contact with the patient or other persons?

A medical device and its nature of contact with patients or other persons has an impact on risks, risk mitigations, safety and performance testing and instructions for use.

What materials or components are utilized in the medical device or are used with, or are in contact with, the medical device?

The materials or components utilized in the medical device should be listed for this question. Certain ingredients may have a particular role or purpose and these attributes should be identified and explained in terms of clinical impact. Primary packaging should also be identified with a description of each component and its purpose of functions.

Is energy delivered to or extracted from the patient?

Battery or mains powered devices in some capacity deliver energy to a patient or extracts energy. Typically, these medical devices incorporate software.

> If this question applies to the medical device, the ways in which energy is delivered and/or extracted should be documented here.

Are substances delivered to or extracted from the patient?

Delivery devices include transfusion pumps, therapeutic patches and other mechanisms that provide pathways for delivery of medications. Extraction can include, catheters, vacutainers or diagnostic devices. The manner and mechanism of delivery and extraction should be described if this question applies to the medical device.

Are biological materials processed by the medical device for subsequent reuse, transfusion or transplantation?

For filtration, sterilization this question may be relevant.

Questions 8-14

Is the medical device supplied sterile or intended to be sterilized by the user, or are other microbiological controls applicable?

The purpose of this question is to prompt information relating to sterility and also impacts the life time of the device e.g. single use, disposable suitable for re-sterilization and re-use. If re-sterilization by the user is permitted, then the manufacturer should note the appropriate steps required or refer to instructions for use that adequately ensure safe re-use.

Is the medical device intended to be routinely cleaned and disinfected by the user?

Devices which can be routinely cleaned, disinfected and re-used by the user can carry additional risks. Instructions for use provide a form of mitigation for such devices. Frequency of cleaning and disinfection are important factors to consider as well as cleaning methods.

Does the medical device modify the patient environment?

The applicability of this question is limited to devices that modify a patients surroundings.

Are measurements taken?

Devices with a measurement function can include blood pressure monitors, urinalysis tests or more complex devices such as ECG machines. The manufacturer should detail the purpose of the measurement system, its operating or measurement principles and how the measurement is read or interpreted.

Is the medical device interpretative?

This refers to if the capability of a device to present the user with a conclusion or conclusions (result, number, pass/fail.

Is the medical device intended for use in conjunction with other medical devices, medicines or other medical technologies?

Identify any accessories or other medical devices or medicines that can be used together with the device. Consideration to potential problems with interactions should be examined.

Are there unwanted outputs of energy or substances?

Depending on the device and the technology used, unwanted energy-related factors e.g. noise, heat, vibration.

Questions 15-21

Is the medical device susceptible to environmental influences?

Does the storage environment affect the device in anyway. Or is there specific transportation requirements or operational requirements that need to be met.

Does the medical device influence the environment?

This refers to any devices that may release or emit waste or dangerous materials. Other factors may include electromagnetic disturbance or noise.

Does the medical device require consumables or accessories?

Consumables or accessories may have specific requirements around their safe use, cleaning, setup, suitability or disposal.

Is maintenance or calibration necessary?

If maintenance is required is this maintenance to be completed out by the device user or a specialist technician.

Does the medical device contain software?

Factors to consider if the device contains software is if the software is intended to be installed, verified, modified or exchanged by the user or by a specialist, and the authenticity of a software update.

Does the medical device allow access to information?

Software based medical devices may allow connectivity in the form of USV devices, or ethernet connections.

Does the medical device store data critical to patient care?

If the device stores critical consideration should be given to the risks of the data being modified or corrupted, unauthorized access and the potential consequences for patients.

Does the medical device have a restricted shelf-life?

This questions refers to factors which may require special storage conditions or effects of storage.

Questions 22-29

Does the medical device have a restricted shelf life?

Factors that should be considered include whether the medical device can deteriorate over time, the impact of storage conditions and primary packaging, the communication of the expiry date (by labelling or an indicator), possibility of use after the expiry date, and the disposal of expired medical devices.

Are there any delayed or long-term use effects?

Factors that should be considered include ergonomic and cumulative effects. Examples could include pumps for saline that corrode over time, mechanical fatigue, loosening of straps and attachments, vibration effects, labels that wear or fall off, long-term material degradation.

To what mechanical forces will the medical device be subjected?

Factors that should be considered include whether the forces to which the medical device will be subjected are under the control of the user or controlled by interaction with other persons.

What determines the lifetime of the medical device?

Factors that should be considered include battery depletion, deterioration of materials and failure of components due to ageing, wear, fatigue or repeated use. The availability of spare parts should be considered as well.

Is the medical device intended for single use?

Factors that should be considered include:

— whether the medical device self-destructs after use;

— whether it is obvious to the user that the medical device has been used

Is safe decommissioning or disposal of the medical device necessary?

Factors that should be considered include the waste products that are generated during the disposal of the medical device itself, and the proper sanitization (removal) of all sensitive data on the medical device. For example, does it contain hazardous material (e.g. toxic chemical or biological agent), or is the material recyclable? If the medical device stores data, proper handling and security of the stored data should be considered, including data removal and retention.

Does installation or use of the medical device require special training or special skills?

Factors that should be considered include the complexity and novelty of the medical device and the knowledge, skills and ability of the persons installing, maintaining or using the medical device. This can include training, education, competence assessment, certification or qualification.

How will information for safety be provided?

Factors that should be considered include:

— whether information will be provided directly to the end user by the manufacturer or will it involve the participation of third parties such as installers, care providers, health care professionals, laboratory directors or pharmacists and whether this will have implications for training;

— commissioning and transferring to the end user and whether it is likely/possible that installation can be carried out by people without the necessary skills;

— based on the type and expected lifetime of the medical device, whether re-training or re-certification of users or service personnel would be indicated.

Questions 30- 31.6

Are new manufacturing processes established or introduced?

Factors that should be considered include the application of new or innovative technology and changes in the scale of production. This can also involve changes in contract manufacturing, suppliers and vendors.

Is successful application of the medical device dependent on the usability of the user interface?

Can the user interface design features contribute to use error?

Factors that should be considered include: control and indicators, symbols used, ergonomic features, physical design and layout, hierarchy of operation, menus for software-driven medical devices, visibility of warnings, audibility of alarms, standardisation of colour coding.

Is the medical device used in an environment where distractions can cause use error?

Factors that should be considered include:

— the consequence of use error;

— whether the distractions are commonplace;

— whether the user can be disturbed by an infrequent distraction;

— whether repetitive stress can reduce the user's awareness or attention.

Does the medical device have connecting parts or accessories?

Factors that should be considered include the possibility of wrong connections, similarity to other products' connections, connection force, feedback on connection integrity, and over- and under tightening.

Does the medical device have a control interface?

Factors that should be considered include spacing, coding, grouping, mapping, modes of feedback, blunders, slips, control differentiation, visibility, direction of activation or change, whether the controls are continuous or discrete, and the reversibility of settings or actions

Does the medical device display information?

Factors that should be considered include visibility in various environments, orientation, the visual capabilities of the user, populations and perspectives, clarity of the presented information, units, colour coding, and the accessibility of critical information.

Is the medical device controlled by a menu?

Factors that should be considered include complexity and number of layers, awareness of state, location of settings, navigation method, number of steps per action, sequence clarity and memorization problems, and importance of control function relative to its accessibility and the impact of deviating from specified operating procedures.

Questions 31.7 -37

Is the successful use of the medical device dependent on a user's knowledge, skills and abilities?

Factors that should be considered include:

— the (intended) users, their mental and physical abilities, skill and training;

— the use environment, ergonomic aspects, installation requirements;

— the capability of intended users to control or influence the use of the medical device; and

the personal characteristics of intended users that can affect their ability to successfully interact with the medical device.

Will the medical device be used by persons with specific needs?

Factors that should be considered include:

— users with special characteristics, such as disabled persons, the elderly and children, who might need assistance by another person to enable the use of a medical device;

— users having wide-ranging skill levels and differing cultural backgrounds and expectations that could lead to differences in what is considered appropriate application of the medical device.

Can the user interface be used to initiate unauthorised actions?

Factors that should be considered include whether the user interface allows the user to enter an operation mode with restricted access (e.g. for maintenance or special use), which increases the possibility of use error and thereby the associated risks, and whether the user becomes aware of having entered such operation mode.

Does the medical device include an alarm system?

Factors that should be considered are the risk of false alarms, missing alarms, disconnected alarm systems, unreliable remote alarm systems, and the user's ability of understanding how the alarm system works.

In what ways might the medical device be misused (deliberately or not)?

> Factors that should be considered are incorrect use of connectors, disabling safety features or alarms, neglect of manufacturer's recommended maintenance, unauthorized access to the medical device or to medical device functions.

Is the medical device intended to be mobile or portable?

> Factors that should be considered are the need for grips, handles, wheels or brakes, and the need for mechanical stability and durability

Does the use of the medical device depend on essential performance?

Factors that should be considered are, for example, the characteristics of the output of life supporting medical devices or the operation of an alarm.

Does the medical device have a degree of autonomy?

Factors that should be considered include:

— awareness of the user when the medical device with a degree of autonomy generates an error, alarm or failure;

— awareness of the user when intervention in an autonomously performed action is required;

— the ability of the user to intervene in or to abort an action that is performed autonomously; and

— the ability of the user to select and perform proper corrective actions.

Does the medical device produce an output that is used as an input in determining clinical action?

Factors that should be considered include whether incorrect or delayed outputs can result in direct or indirect risks to patients, e.g. an incorrect diagnosis resulting in delayed or omitted therapy for a patient

APPENDIX I- Use Engineering and Product Development

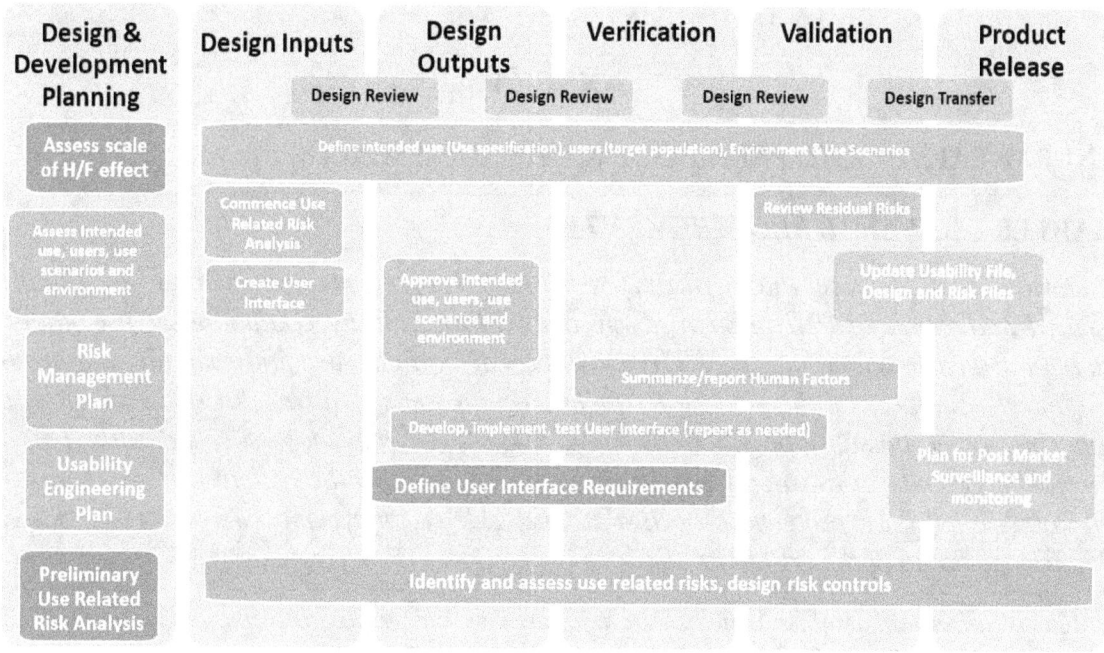

APPENDIX MDR Annex I, General Safety and Performance Requirements

ANNEX I

GENERAL SAFETY AND PERFORMANCE REQUIREMENTS

CHAPTER I *GENERAL REQUIREMENTS*

1. Devices shall achieve the performance intended by their manufacturer and shall be designed and manufactured in such a way that, during normal conditions of use, they are suitable for their intended purpose. They shall be safe and effective and shall not compromise the clinical condition or the safety of patients, or the safety and health of users or, where applicable, other persons, provided that any risks which may be associated with their use constitute acceptable risks when weighed against the benefits to the patient and are compatible with a high level of protection of health and safety, taking into account the generally acknowledged state of the art.

2. The requirement in this Annex to reduce risks as far as possible means the reduction of risks as far as possible without adversely affecting the benefit-risk ratio.

3. Manufacturers shall establish, implement, document and maintain a risk management system.

Risk management shall be understood as a continuous iterative process throughout the entire lifecycle of a device, requiring regular systematic updating. In carrying out risk management manufacturers shall:

(a) establish and document a risk management plan for each device;

(b) identify and analyse the known and foreseeable hazards associated with each device;

(c) estimate and evaluate the risks associated with, and occurring during, the intended use and during reasonably foreseeable misuse;

(d) eliminate or control the risks referred to in point (c) in accordance with the requirements of Section 4;

(e) evaluate the impact of information from the production phase and, in particular, from the post-market surveillance system, on hazards and the frequency of occurrence thereof, on

estimates of their associated risks, as well as on the overall risk, benefit-risk ratio and risk acceptability; and

(f) based on the evaluation of the impact of the information referred to in point (e), if necessary amend control measures in line with the requirements of Section 4.

4. Risk control measures adopted by manufacturers for the design and manufacture of the devices shall conform to safety principles, taking account of the generally acknowledged state of the art. To reduce risks, Manufacturers shall manage risks so that the residual risk associated with each hazard as well as the overall residual risk is judged acceptable. In selecting the most appropriate solutions, manufacturers shall, in the following order of priority:

(a) eliminate or reduce risks as far as possible through safe design and manufacture;

(b) where appropriate, take adequate protection measures, including alarms if necessary, in relation to risks that cannot be eliminated; and

(c) provide information for safety (warnings/precautions/contra-indications) and, where appropriate, training to users.

Manufacturers shall inform users of any residual risks.

5. In eliminating or reducing risks related to use error, the manufacturer shall:

(a) reduce as far as possible the risks related to the ergonomic features of the device and the environment in which the device is intended to be used (design for patient safety), and

(b) give consideration to the technical knowledge, experience, education, training and use environment, where applicable, and the medical and physical conditions of intended users (design for lay, professional, disabled or other users).

6. The characteristics and performance of a device shall not be adversely affected to such a degree that the health or safety of the patient or the user and, where applicable, of other persons are compromised during the lifetime of the device, as indicated by the manufacturer, when the device is subjected to the stresses which can occur during normal conditions of use and has been properly maintained in accordance with the manufacturer's instructions.

7. Devices shall be designed, manufactured and packaged in such a way that their characteristics and performance during their intended use are not adversely affected during transport and storage, for example, through fluctuations of temperature and humidity, taking account of the instructions and information provided by the manufacturer.

8. All known and foreseeable risks, and any undesirable side-effects, shall be minimised and be acceptable when weighed against the evaluated benefits to the patient and/or user arising from the achieved performance of the device during normal conditions of use.

9. For the devices referred to in Annex XVI, the general safety requirements set out in Sections 1 and 8 shall be understood to mean that the device, when used under the conditions and for the purposes intended, does not present a risk at all or presents a risk that is no more than the maximum acceptable risk related to the product's use which is consistent with a high level of protection for the safety and health of persons.

CHAPTER II REQUIREMENTS REGARDING DESIGN AND MANUFACTURE

10. Chemical, physical and biological properties

10.1. Devices shall be designed and manufactured in such a way as to ensure that the characteristics and performance requirements referred to in Chapter I are fulfilled. Particular attention shall be paid to:

(a) the choice of materials and substances used, particularly as regards toxicity and, where relevant, flammability;

(b) the compatibility between the materials and substances used and biological tissues, cells and body fluids, taking account of the intended purpose of the device and, where relevant, absorption, distribution, metabolism and excretion;

(c) the compatibility between the different parts of a device which consists of more than one implantable part;

(d) the impact of processes on material properties;

(e) where appropriate, the results of biophysical or modelling research the validity of which has been demonstrated beforehand;

(f) the mechanical properties of the materials used, reflecting, where appropriate, matters such as strength, ductility, fracture resistance, wear resistance and fatigue resistance;

(g) surface properties; and

(h) the confirmation that the device meets any defined chemical and/or physical specifications.

10.2. Devices shall be designed, manufactured and packaged in such a way as to minimise the risk posed by contaminants and residues to patients, taking account of the intended purpose of the device, and to the persons involved in the transport, storage and use of the devices. Particular attention shall be paid to tissues exposed to those contaminants and residues and to the duration and frequency of exposure.

10.3. Devices shall be designed and manufactured in such a way that they can be used safely with the materials and substances, including gases, with which they enter into contact during their intended use; if the devices are intended to administer medicinal products they shall be designed and manufactured in such a way as to be compatible with the medicinal products concerned in accordance with the provisions and restrictions governing those medicinal products and that the performance of both the medicinal products and of the devices is maintained in accordance with their respective indications and intended use.

10.4. Substances

10.4.1. Design and manufacture of devices

Devices shall be designed and manufactured in such a way as to reduce as far as possible the risks posed by substances or particles, including wear debris, degradation products and processing residues, that may be released from the device.

Devices, or those parts thereof or those materials used therein that:

— are invasive and come into direct contact with the human body,

— (re)administer medicines, body liquids or other substances, including gases, to/from the body, or

— transport or store such medicines, body fluids or substances, including gases, to be (re)administered to the body,

shall only contain the following substances in a concentration that is above 0,1 % weight by weight (w/w) where justified pursuant to Section 10.4.2:

(a) substances which are carcinogenic, mutagenic or toxic to reproduction ('CMR'), of category 1A or 1B, in accordance with Part 3 of Annex VI to Regulation (EC) No 1272/2008 of the European Parliament and of the Council ( ), or

(b) substances having endocrine-disrupting properties for which there is scientific evidence of probable serious effects to human health and which are identified either in accordance with the procedure set out in Article 59 of Regulation (EC) No 1907/2006 of the European Parliament and of the Council (²) or, once a delegated act has been adopted by the Commission pursuant to the first subparagraph of Article 5(3) of Regulation (EU) No 528/2012 of the European Parliament and the Council (³), in accordance with the criteria that are relevant to human health amongst the criteria established therein.

10.4.2. Justification regarding the presence of CMR and/or endocrine-disrupting substances

The justification for the presence of such substances shall be based upon:

(a) an analysis and estimation of potential patient or user exposure to the substance;

(b) an analysis of possible alternative substances, materials or designs, including, where available, information about independent research, peer-reviewed studies, scientific opinions from relevant scientific committees and an analysis of the availability of such alternatives;

(c) argumentation as to why possible substance and/ or material substitutes, if available, or design changes, if feasible, are inappropriate in relation to maintaining the functionality, performance and the benefit-risk ratios of the product; including taking into account if the intended use of such devices includes treatment of children or treatment of pregnant or breastfeeding women or treatment of other patient groups considered particularly vulnerable to such substances and/or materials; and

(d) where applicable and available, the latest relevant scientific committee guidelines in accordance with Sections 10.4.3. and 10.4.4.

10.4.3. Guidelines on phthalates

For the purposes of Section 10.4., the Commission shall, as soon as possible and by 26 May 2018, provide the relevant scientific committee with a mandate to prepare guidelines that shall be ready before 26 May 2020. The mandate for the committee shall encompass at least a benefit-risk assessment of the presence of phthalates which belong to either of the groups of substances referred to in points (a) and (b) of Section 10.4.1. The benefit-risk assessment shall take into account the intended purpose and context of the use of the device, as well as any available alternative substances and alternative materials, designs or medical treatments. When *deemed appropriate on the basis of the latest scientific evidence, but at least every five years, the guidelines shall be updated.*

10.4.4. Guidelines on other CMR and endocrine-disrupting substances

Subsequently, the Commission shall mandate the relevant scientific committee to prepare guidelines as referred to in Section 10.4.3. also for other substances referred to in points (a) and (b) of Section 10.4.1., where appropriate.

10.4.5. Labelling

Where devices, parts thereof or materials used therein as referred to in Section 10.4.1. contain substances referred to in points (a) or (b) of Section 10.4.1. in a concentration above 0,1 % weight by weight (w/w), the presence of those substances shall be labelled on the device itself and/or on the packaging for each unit or, where appropriate, on the sales packaging, with the list of such substances. If the intended use of such devices includes treatment of children or treatment of pregnant or breastfeeding women or treatment of other patient groups considered particularly vulnerable to such substances and/or materials, information on residual risks for those patient groups and, if applicable, on appropriate precautionary measures shall be given in the instructions for use.

10.5. Devices shall be designed and manufactured in such a way as to reduce as far as possible the risks posed by the unintentional ingress of substances into the device taking into account the device and the nature of the environment in which it is intended to be used.

10.6. Devices shall be designed and manufactured in such a way as to reduce as far as possible the risks linked to the size and the properties of particles which are or can be released into the patient's or user's body, unless they come into contact with intact skin only. Special attention shall be given to nanomaterials.

11. Infection and microbial contamination

11.1. Devices and their manufacturing processes shall be designed in such a way as to eliminate or to reduce as far as possible the risk of infection to patients, users and, where applicable, other persons. The design shall:

(a) reduce as far as possible and appropriate the risks from unintended cuts and pricks, such as needle stick injuries,

(b) allow easy and safe handling,

(c) reduce as far as possible any microbial leakage from the device and/or microbial exposure during use, and

(d) prevent microbial contamination of the device or its content such as specimens or fluids.

11.2. Where necessary devices shall be designed to facilitate their safe cleaning, disinfection, and/or re-sterilisation.

11.3. Devices labelled as having a specific microbial state shall be designed, manufactured and packaged to ensure that they remain in that state when placed on the market and remain so under the transport and storage conditions specified by the manufacturer.

11.4. Devices delivered in a sterile state shall be designed, manufactured and packaged in accordance with appropriate procedures, to ensure that they are sterile when placed on the market and that, unless the packaging which is intended to maintain their sterile condition is damaged, they remain sterile, under the transport and storage conditions specified by the manufacturer, until that packaging is opened at the point of use. It shall be ensured that the integrity of that packaging is clearly evident to the final user.

11.5. Devices labelled as sterile shall be processed, manufactured, packaged and, sterilised by means of appropriate, validated methods.

11.6. Devices intended to be sterilised shall be manufactured and packaged in appropriate and controlled conditions and facilities.

11.7. Packaging systems for non-sterile devices shall maintain the integrity and cleanliness of the product and, where the devices are to be sterilised prior to use, minimise the risk of microbial contamination; the packaging system shall be suitable taking account of the method of sterilisation indicated by the manufacturer.

11.8. The labelling of the device shall distinguish between identical or similar devices placed on the market in both a sterile and a non-sterile condition additional to the symbol used to indicate that devices are sterile.

12. Devices incorporating a substance considered to be a medicinal product and devices that are composed of substances or of combinations of substances that are absorbed by or locally dispersed in the human body.

12.1. In the case of devices referred to in the first subparagraph of Article 1(8), the quality, safety and usefulness of the substance which, if used separately, would be considered to be a medicinal product within the meaning of point (2) of Article 1 of Directive 2001/83/EC, shall *be verified by analogy with the methods specified in Annex I to Directive 2001/83/EC, as required by the applicable conformity assessment procedure under this Regulation.*

12.2. Devices that are composed of substances or of combinations of substances that are intended to be introduced into the human body, and that are absorbed by or locally dispersed in the human body shall comply, where applicable and in a manner limited to the aspects not covered by this Regulation, with the relevant requirements laid down in Annex I to Directive 2001/83/EC for the evaluation of absorption, distribution, metabolism, excretion, local tolerance, toxicity, interaction with other devices, medicinal products or other substances and potential for adverse reactions, as required by the applicable conformity assessment procedure under this Regulation.

13. Devices incorporating materials of biological origin

13.1. For devices manufactured utilising derivatives of tissues or cells of human origin which are non-viable or are rendered non-viable covered by this Regulation in accordance with point (g) of Article 1(6), the following shall apply:

(a) donation, procurement and testing of the tissues and cells shall be done in accordance with Directive 2004/23/EC;

(b) processing, preservation and any other handling of those tissues and cells or their derivatives shall be carried out so as to provide safety for patients, users and, where applicable, other persons. In particular, safety with regard to viruses and other transmissible agents shall be addressed by appropriate methods of sourcing and by implementation of validated methods of elimination or inactivation in the course of the manufacturing process;

(c) the traceability system for those devices shall be complementary and compatible with the traceability and data protection requirements laid down in Directive 2004/23/EC and in Directive 2002/98/EC.

13.2. For devices manufactured utilising tissues or cells of animal origin, or their derivatives, which are non-viable or rendered non-viable the following shall apply:

(a) where feasible taking into account the animal species, tissues and cells of animal origin, or their derivatives, shall originate from animals that have been subjected to veterinary controls that are adapted to the intended use of the tissues. Information on the geographical origin of the animals shall be retained by manufacturers;

(b) sourcing, processing, preservation, testing and handling of tissues, cells and substances of

animal origin, or their derivatives, shall be carried out so as to provide safety for patients, users and, where applicable, other persons. In particular safety with regard to viruses and other transmissible agents shall be addressed by implementation of validated methods of elimination or viral inactivation in the course of the manufacturing process, except when the use of such methods would lead to unacceptable degradation compromising the clinical benefit of the device;

(c) in the case of devices manufactured utilising tissues or cells of animal origin, or their derivatives, as referred to in Regulation (EU) No 722/2012 the particular requirements laid down in that Regulation shall apply.

13.3. For devices manufactured utilising non-viable biological substances other than those referred to in Sections 13.1 and 13.2, the processing, preservation, testing and handling of those substances shall be carried out so as to provide safety for patients, users and, where applicable, other persons, including in the waste disposal chain. In particular, safety with regard to viruses and other transmissible agents shall be addressed by appropriate methods of sourcing and by implementation of validated methods of elimination or inactivation in the course of the manufacturing process.

14. Construction of devices and interaction with their environment

14.1. If the device is intended for use in combination with other devices or equipment the whole combination, including the connection system shall be safe and shall not impair the specified performance of the devices. Any restrictions on use applying to such combinations shall be indicated on the label and/or in the instructions for use. Connections which the user has to handle, such as fluid, gas transfer, electrical or mechanical coupling, shall be designed and constructed in such a way as to minimise all possible risks, such as misconnection.

14.2. Devices shall be designed and manufactured in such a way as to remove or reduce as far as possible:

(a) the risk of injury, in connection with their physical features, including the volume/pressure ratio, dimensional and where appropriate ergonomic features;

(b) risks connected with reasonably foreseeable external influences or environmental conditions, such as magnetic fields, external electrical and electromagnetic effects, electrostatic discharge, radiation associated with diagnostic or therapeutic procedures, pressure, humidity, temperature, variations in pressure and acceleration or radio signal interferences;

(c) the risks associated with the use of the device when it comes into contact with materials, liquids, and substances, including gases, to which it is exposed during normal conditions of use;

(d) the risks associated with the possible negative interaction between software and the IT environment within which it operates and interacts;

(e) the risks of accidental ingress of substances into the device;

(f) the risks of reciprocal interference with other devices normally used in the investigations or for the treatment given; and

(g) risks arising where maintenance or calibration are not possible (as with implants), from ageing of materials used or loss of accuracy of any measuring or control mechanism.

14.3. Devices shall be designed and manufactured in such a way as to minimise the risks of fire or explosion during normal use and in single fault condition. Particular attention shall be paid to devices the intended use of which includes exposure to or use in association with flammable or explosive substances or substances which could cause combustion.

14.4. Devices shall be designed and manufactured in such a way that adjustment, calibration, and maintenance can be done safely and effectively.

14.5. Devices that are intended to be operated together with other devices or products shall be designed and manufactured in such a way that the interoperability and compatibility are reliable and safe.

14.6 Any measurement, monitoring or display scale shall be designed and manufactured in line with ergonomic principles, taking account of the intended purpose, users and the environmental conditions in which the devices are intended to be used.

14.7. Devices shall be designed and manufactured in such a way as to facilitate their safe disposal and the safe disposal of related waste substances by the user, patient or other person. To that end, manufacturers shall identify and test procedures and measures as a result of which their devices can be safely disposed after use. Such procedures shall be described in the instructions for use.

15. Devices with a diagnostic or measuring function

15.1. Diagnostic devices and devices with a measuring function, shall be designed and manufactured in such a way as to provide sufficient accuracy, precision and stability for their intended purpose, based on appropriate scientific and technical methods. The limits of accuracy shall be indicated by the manufacturer.

15.2. The measurements made by devices with a measuring function shall be expressed in legal units conforming to the provisions of Council Directive 80/181/EEC (¹).

16. Protection against radiation

16.1. General

(a) Devices shall be designed, manufactured and packaged in such a way that exposure of patients, users and other persons to radiation is reduced as far as possible, and in a manner that is compatible with the intended purpose, whilst not restricting the application of appropriate specified levels for therapeutic and diagnostic purposes.

(b) The operating instructions for devices emitting hazardous or potentially hazardous radiation shall contain detailed information as to the nature of the emitted radiation, the means of protecting the patient and the user, and on ways of avoiding misuse and of reducing the risks inherent to installation as far as possible and appropriate. Information regarding the acceptance and performance testing, the acceptance criteria, and the maintenance procedure shall also be specified.

16.2. Intended radiation

(a) Where devices are designed to emit hazardous, or potentially hazardous, levels of ionizing and/or non-ionizing radiation necessary for a specific medical purpose the benefit of which is considered to outweigh the risks inherent to the emission, it shall be possible for the user to control the emissions. Such devices shall be designed and manufactured to ensure reproducibility of relevant variable parameters within an acceptable tolerance.

(b) Where devices are intended to emit hazardous, or potentially hazardous, ionizing and/or non-ionizing radiation, they shall be fitted, where possible, with visual displays and/or audible warnings of such emissions.

16.3. Devices shall be designed and manufactured in such a way that exposure of patients, users and other persons to the emission of unintended, stray or scattered radiation is reduced as far as possible. Where possible and appropriate, methods shall be selected which reduce the exposure to radiation of patients, users and other persons who may be affected.

16.4. Ionising radiation

(a) Devices intended to emit ionizing radiation shall be designed and manufactured taking into account the requirements of the Directive 2013/59/Euratom laying down basic safety standards for protection against the dangers arising from exposure to ionising radiation.

(b) Devices intended to emit ionising radiation shall be designed and manufactured in such a way as to ensure that, where possible, taking into account the intended use, the quantity, geometry and quality of the radiation emitted can be varied and controlled, and, if possible, monitored during treatment.

(c) Devices emitting ionising radiation intended for diagnostic radiology shall be designed and manufactured in such a way as to achieve an image and/or output quality that are appropriate to the intended medical purpose whilst minimising radiation exposure of the patient and user.

(d) Devices that emit ionising radiation and are intended for therapeutic radiology shall be designed and manufactured in such a way as to enable reliable monitoring and control of the delivered dose, the beam type, energy and, where appropriate, the quality of radiation.

17. Electronic programmable systems – devices that incorporate electronic programmable systems and software that are devices in themselves

17.1. Devices that incorporate electronic programmable systems, including software, or software that are devices in themselves, shall be designed to ensure repeatability, reliability and performance in line with their intended use. In the event of a single fault condition, appropriate means shall be adopted to eliminate or reduce as far as possible consequent risks or impairment of performance.

17.2. For devices that incorporate software or for software that are devices in themselves, the software shall be developed and manufactured in accordance with the state of the art taking into account the principles of development life cycle, risk management, including information security, verification and validation.

17.3. Software referred to in this Section that is intended to be used in combination with mobile computing platforms shall be designed and manufactured taking into account the specific features of the mobile platform (e.g. size and contrast ratio of the screen) and the external factors related to their use (varying environment as regards level of light or noise).

17.4. Manufacturers shall set out minimum requirements concerning hardware, IT networks characteristics and IT security measures, including protection against unauthorised access, necessary to run the software as intended.

18. Active devices and devices connected to them

18.1. For non-implantable active devices, in the event of a single fault condition, appropriate means shall be adopted to eliminate or reduce as far as possible consequent risks.

18.2. Devices where the safety of the patient depends on an internal power supply shall be equipped with a means of determining the state of the power supply and an appropriate warning or indication for when the capacity of the power supply becomes critical. If necessary, such warning or indication shall be given prior to the power supply becoming critical.

18.3. Devices where the safety of the patient depends on an external power supply shall include an alarm system to signal any power failure.

18.4. Devices intended to monitor one or more clinical parameters of a patient shall be equipped with appropriate alarm systems to alert the user of situations which could lead to death or severe deterioration of the patient's state of health.

18.5. Devices shall be designed and manufactured in such a way as to reduce as far as possible the risks of creating electromagnetic interference which could impair the operation of the device in question or other devices or equipment in the intended environment.

18.6. Devices shall be designed and manufactured in such a way as to provide a level of intrinsic immunity to electromagnetic interference such that is adequate to enable them to operate as intended.

18.7. Devices shall be designed and manufactured in such a way as to avoid, as far as possible, the risk of accidental electric shocks to the patient, user or any other person, both during normal use of the device and in the event of a single fault condition in the device, provided the device is installed and maintained as indicated by the manufacturer.

18.8. Devices shall be designed and manufactured in such a way as to protect, as far as possible, against unauthorised access that could hamper the device from functioning as intended.

19. Particular requirements for active implantable devices

19.1. Active implantable devices shall be designed and manufactured in such a way as to remove or minimize as far as possible:

(a) risks connected with the use of energy sources with particular reference, where electricity is used, to insulation, leakage currents and overheating of the devices,

(b) risks connected with medical treatment, in particular those resulting from the use of defibrillators or high-frequency surgical equipment, and

(c) risks which may arise where maintenance and calibration are impossible, including:

- *excessive increase of leakage currents,*

- *ageing of the materials used,*

- *excess heat generated by the device,*

- *decreased accuracy of any measuring or control mechanism.*

19.2. Active implantable devices shall be designed and manufactured in such a way as to ensure

– if applicable, the compatibility of the devices with the substances they are intended to administer, and

- *the reliability of the source of energy.*

19.3. Active implantable devices and, if appropriate, their component parts shall be identifiable to allow any necessary measure to be taken following the discovery of a potential risk in connection with the devices or their component parts.

19.4. Active implantable devices shall bear a code by which they and their manufacturer can be unequivocally identified (particularly with regard to the type of device and its year of manufacture); it shall be possible to read this code, if necessary, without the need for a surgical operation.

20. Protection against mechanical and thermal risks

20.1. Devices shall be designed and manufactured in such a way as to protect patients and users against mechanical risks connected with, for example, resistance to movement, instability and moving parts.

20.2. Devices shall be designed and manufactured in such a way as to reduce to the lowest possible level the risks arising from vibration generated by the devices, taking account of technical progress and of the means available for limiting vibrations, particularly at source, unless the vibrations are part of the specified performance.

20.3. Devices shall be designed and manufactured in such a way as to reduce to the lowest possible level the risks arising from the noise emitted, taking account of technical progress and of the means available to reduce noise, particularly at source, unless the noise emitted is part of the specified performance.

20.4. Terminals and connectors to the electricity, gas or hydraulic and pneumatic energy supplies which the user or other person has to handle, shall be designed and constructed in such a way as to minimise all possible risks.

20.5. Errors likely to be made when fitting or refitting certain parts which could be a source of risk shall be made impossible by the design and construction of such parts or, failing this, by information given on the parts themselves and/or their housings.

The same information shall be given on moving parts and/or their housings where the direction of movement needs to be known in order to avoid a risk.

20.6. Accessible parts of devices (excluding the parts or areas intended to supply heat or reach given temperatures) and their surroundings shall not attain potentially dangerous temperatures under normal conditions of use.

21. Protection against the risks posed to the patient or user by devices supplying energy or substances

21.1. Devices for supplying the patient with energy or substances shall be designed and constructed in such a way that the amount to be delivered can be set and maintained accurately enough to ensure the safety of the patient and of the user.

21.2. Devices shall be fitted with the means of preventing and/or indicating any inadequacies in the amount of energy delivered or substances delivered which could pose a danger. Devices shall incorporate suitable means to prevent, as far as possible, the accidental release of dangerous levels of energy or substances from an energy and/or substance source.

21.3. The function of the controls and indicators shall be clearly specified on the devices. Where a device bears instructions required for its operation or indicates operating or adjustment parameters by means of a visual system, such information shall be understandable to the user and, as appropriate, the patient.

22. Protection against the risks posed by medical devices intended by the manufacturer for use by lay persons

22.1. Devices for use by lay persons shall be designed and manufactured in such a way that they perform appropriately for their intended purpose taking into account the skills and the means available to lay persons and the influence resulting from variation that can be reasonably anticipated in the lay person's technique and environment. The information and instructions provided by the manufacturer shall be easy for the lay person to understand and apply.

22.2. Devices for use by lay persons shall be designed and manufactured in such a way as to:

— ensure that the device can be used safely and accurately by the intended user at all stages of the procedure, if necessary after appropriate training and/or information,

— reduce, as far as possible and appropriate, the risk from unintended cuts and pricks such as needle stick injuries, and

— reduce as far as possible the risk of error by the intended user in the handling of the device and, if applicable, in the interpretation of the results.

22.3. Devices for use by lay persons shall, where appropriate, include a procedure by which the lay person:

— can verify that, at the time of use, the device will perform as intended by the manufacturer, and

— if applicable, is warned if the device has failed to provide a valid result.

CHAPTER III

REQUIREMENTS REGARDING THE INFORMATION SUPPLIED WITH THE DEVICE

23. Label and instructions for use

23.1. General requirements regarding the information supplied by the manufacturer

Each device shall be accompanied by the information needed to identify the device and its manufacturer, and by any safety and performance information relevant to the user, or any other person, as appropriate. Such information may appear on the device itself, on the packaging or in the instructions for use, and shall, if the manufacturer has a website, be made available and kept up to date on the website, taking into account the following:

(a) The medium, format, content, legibility, and location of the label and instructions for use shall be appropriate to the particular device, its intended purpose and the technical knowledge, experience, education or training of the intended user(s). In particular, instructions for use shall be written in terms readily understood by the intended user and, where appropriate, supplemented with drawings and diagrams.

(b) The information required on the label shall be provided on the device itself. If this is not practicable or appropriate, some or all of the information may appear on the packaging for each unit, and/or on the packaging of multiple devices.

(c) Labels shall be provided in a human-readable format and may be supplemented by machine-readable information, such as radio-frequency identification ('RFID') or bar codes.

(d) Instructions for use shall be provided together with devices. By way of exception, instructions for use shall not be required for class I and class IIa devices if such devices can be used safely without any such instructions and unless otherwise provided for elsewhere in this Section.

(e) Where multiple devices are supplied to a single user and/or location, a single copy of the instructions for use may be provided if so agreed by the purchaser who in any case may request further copies to be provided free of charge.

(f) Instructions for use may be provided to the user in non-paper format (e.g. electronic) to the extent, and only under the conditions, set out in Regulation (EU) No 207/2012 or in any subsequent implementing rules adopted pursuant to this Regulation.

(g) Residual risks which are required to be communicated to the user and/or other person shall be included as limitations, contra-indications, precautions or warnings in the information supplied by the manufacturer.

(h) Where appropriate, the information supplied by the manufacturer shall take the form of internationally recognised symbols. Any symbol or identification colour used shall conform to the harmonised standards or CS. In areas for which no harmonised standards or CS exist, the symbols and colours shall be described in the documentation supplied with the device.

23.2. Information on the label

The label shall bear all of the following particulars:

(a) the name or trade name of the device;

(b) the details strictly necessary for a user to identify the device, the contents of the packaging and, where it is not obvious for the user, the intended purpose of the device;

(c) the name, registered trade name or registered trade mark of the manufacturer and the address of its registered place of business;

(d) if the manufacturer has its registered place of business outside the Union, the name of the authorised representative and address of the registered place of business of the authorised representative;

(e) where applicable, an indication that the device contains or incorporates:

   — a medicinal substance, including a human blood or plasma derivative, or

   — tissues or cells, or their derivatives, of human origin, or

   — tissues or cells of animal origin, or their derivatives, as referred to in Regulation (EU) No 722/2012;

(f) where applicable, information labelled in accordance with Section 10.4.5.;

(g) the lot number or the serial number of the device preceded by the words LOT NUMBER or SERIAL NUMBER or an equivalent symbol, as appropriate;

(h) the UDI carrier referred to in Article 27(4) and Part C of Annex VII;

(i) an unambiguous indication of t the time limit for using or implanting the device safely, expressed at least in terms of year and month, where this is relevant;

(j) where there is no indication of the date until when it may be used safely, the date of manufacture. This date of manufacture may be included as part of the lot number or serial number, provided the date is clearly identifiable;

(k) an indication of any special storage and/or handling condition that applies;

(l) if the device is supplied sterile, an indication of its sterile state and the sterilisation method;

(m) warnings or precautions to be taken that need to be brought to the immediate attention of the user of the device, and to any other person. This information may be kept to a minimum in which case more detailed information shall appear in the instructions for use, taking into account the intended users;

(n) if the device is intended for single use, an indication of that fact. A manufacturer's indication of single use shall be consistent across the Union;

(o) if the device is a single-use device that has been reprocessed, an indication of that fact, the number of reprocessing cycles already performed, and any limitation as regards the number of reprocessing cycles;

(p) if the device is custom-made, the words 'custom-made device';

(q) an indication that the device is a medical device. If the device is intended for clinical investigation only, the words 'exclusively for clinical investigation';

(r) in the case of devices that are composed of substances or of combinations of substances that are intended to be introduced into the human body via a body orifice or applied to the skin and that are absorbed by or locally dispersed in the human body, the overall qualitative composition of the device and quantitative information on the main constituent or constituents responsible for achieving the principal intended action;

(s) for active implantable devices, the serial number, and for other implantable devices, the serial number or the lot number.

23.3. Information on the packaging which maintains the sterile condition of a device ('sterile packaging')

The following particulars shall appear on the sterile packaging:

(a) an indication permitting the sterile packaging to be recognised as such,

(b) a declaration that the device is in a sterile condition,

(c) the method of sterilisation,

(d) the name and address of the manufacturer,

(e) a description of the device,

(f) if the device is intended for clinical investigations, the words 'exclusively for clinical investigations',

(g) if the device is custom-made, the words 'custom-made device',

(h) the month and year of manufacture,

(i) an unambiguous indication of the time limit for using or implanting the device safely expressed at least in terms of year and month, and

(j) an instruction to check the instructions for use for what to do if the sterile packaging is damaged or unintentionally opened before use.

23.4. Information in the instructions for use

The instructions for use shall contain all of the following particulars:

(a) the particulars referred to in points (a), (c), (e), (f), (k), (l), (n) and (r) of Section 23.2;

(b) the device's intended purpose with a clear specification of indications, contra-indications, the patient target group or groups, and of the intended users, as appropriate;

(c) where applicable, a specification of the clinical benefits to be expected.

(d) where applicable, links to the summary of safety and clinical performance referred to in Article 32;

(e) the performance characteristics of the device;

(f) where applicable, information allowing the healthcare professional to verify if the device is suitable and select the corresponding software and accessories;

(g) any residual risks, contra-indications and any undesirable side-effects, including information to be conveyed to the patient in this regard;

(h) specifications the user requires to use the device appropriately, e.g. if the device has a measuring function, the degree of accuracy claimed for it;

(i) details of any preparatory treatment or handling of the device before it is ready for use or during its use, such as sterilisation, final assembly, calibration, etc., including the levels of disinfection required to ensure patient safety and all available methods for achieving those levels of disinfection;

(j) any requirements for special facilities, or special training, or particular qualifications of the device user and/or other persons;

(k) the information needed to verify whether the device is properly installed and is ready to perform safely and as intended by the manufacturer, together with, where relevant:

— details of the nature, and frequency, of preventive and regular maintenance, and of any preparatory cleaning or disinfection,

— identification of any consumable components and how to replace them,

— information on any necessary calibration to ensure that the device operates properly and safely during its intended lifetime, and

— methods for eliminating the risks encountered by persons involved in installing, calibrating or servicing devices;

(l) if the device is supplied sterile, instructions in the event of the sterile packaging being damaged or unintentionally opened before use;

(m) if the device is supplied non-sterile with the intention that it is sterilised before use, the appropriate instructions for sterilisation;

(n) if the device is reusable, information on the appropriate processes for allowing reuse, including cleaning, disinfection, packaging and, where appropriate, the validated method of re-sterilisation appropriate to the Member State or Member States in which the device has been placed on the market. Information shall be provided to identify when the device should no longer be reused, e.g. signs of material degradation or the maximum number of allowable reuses;

(o) an indication, if appropriate, that a device can be reused only if it is reconditioned under the responsibility of the manufacturer to comply with the general safety and performance requirements;

(p) if the device bears an indication that it is for single use, information on known characteristics and technical factors known to the manufacturer that could pose a risk if the device were to be re-used. This information shall be based on a specific section of the manufacturer's risk management documentation, where such characteristics and technical factors shall be addressed in detail. If in accordance with point (d) of Section 23.1. no instructions for use are required, this information shall be made available to the user upon request;

(q) for devices intended for use together with other devices and/or general purpose equipment:

—information to identify such devices or equipment, in order to obtain a safe combination, and/or

—information on any known restrictions to combinations of devices and equipment;

(r) if the device emits radiation for medical purposes:

—detailed information as to the nature, type and where appropriate, the intensity and distribution of the emitted radiation,

—the means of protecting the patient, user, or other person from unintended radiation during use of the device;

(s) information that allows the user and/or patient to be informed of any warnings, precautions, contra-indications, measures to be taken and limitations of use regarding the device. That information shall, where relevant, allow the user to brief the patient about any warnings, precautions, contra-indications, measures to be taken and limitations of use regarding the device. The information shall cover, where appropriate:

—warnings, precautions and/or measures to be taken in the event of malfunction of the device or changes in its performance that may affect safety,

—warnings, precautions and/or measures to be taken as regards the exposure to reasonably foreseeable external influences or environmental conditions, such as magnetic fields, external electrical and electromagnetic effects, electrostatic discharge, radiation associated *with diagnostic or therapeutic procedures, pressure, humidity, or temperature,*

—*warnings, precautions and/or measures to be taken as regards the risks of interference posed by the reasonably foreseeable presence of the device during specific diagnostic investigations, evaluations, or therapeutic treatment or other procedures such as electromagnetic interference emitted by the device affecting other equipment,*

—*if the device is intended to administer medicinal products, tissues or cells of human or animal origin, or their derivatives, or biological substances, any limitations or incompatibility in the choice of substances to be delivered,*

—*warnings, precautions and/or limitations related to the medicinal substance or biological material that is incorporated into the device as an integral part of the device; and*

—*precautions related to materials incorporated into the device that contain or consist of CMR substances or endocrine-disrupting substances, or that could result in sensitisation or an allergic reaction by the patient or user;*

(t) in the case of devices that are composed of substances or of combinations of substances that are intended to be introduced into the human body and that are absorbed by or locally dispersed in the human body, warnings and precautions, where appropriate, related to the general profile of interaction of the device and its products of metabolism with other devices, medicinal products and other substances as well as contra-indications, undesirable side-effects and risks relating to overdose;

(u) in the case of implantable devices, the overall qualitative and quantitative information on the materials and substances to which patients can be exposed;

(v) warnings or precautions to be taken in order to facilitate the safe disposal of the device, its accessories and the consumables used with it, if any. This information shall cover, where appropriate:

—*infection or microbial hazards such as explants, needles or surgical equipment contaminated with potentially infectious substances of human origin, and*

— *physical hazards such as from sharps.*

If in accordance with the point (d) of Section 23.1 no instructions for use are required, this information shall be made available to the user upon request;

(w) for devices intended for use by lay persons, the circumstances in which the user should consult a healthcare professional;

(x) for the devices covered by this Regulation pursuant to Article 1(2), information regarding the absence of a clinical benefit and the risks related to use of the device;

(y) date of issue of the instructions for use or, if they have been revised, date of issue and identifier of the latest revision of the instructions for use;

(z) a notice to the user and/or patient that any serious incident that has occurred in relation to the device should be reported to the manufacturer and the competent authority of the Member State in which the user and/or patient is established;

(aa) information to be supplied to the patient with an implanted device in accordance with Article 18;

(ab) for devices that incorporate electronic programmable systems, including software, or software that are devices in themselves, minimum requirements concerning hardware, IT networks characteristics and IT security measures, including protection against unauthorised access, necessary to run the software as intended.

APPENDIX - Risk Management Plan

Notes to Author

1. *Blue italic text is for general guidance purposes. Delete for document approval*
2. *Black italic text is text can should be edited for specific risk management plans, as required*
3. *When document is created locally, pagination should be applied*
4. *An approved template should form the basis of a risk management plan. The template should be revision controlled and each page should identify the document title and document ID.*

Risk Management Plan

Document Title:	Risk Management Plan for *Advance 101, Battery Powered Digital Blood Pressure Monitor, UA1101*
Document ID:	*R7100-21*
Revision:	*A*
Issued	*30 Feb 2022*

Notes of Scope of Activities

1. *Risk management plans must cover the full lifecycle of the product, starting with product development to post launch product lifecycle risk management.*
2. *The extent of planned activities and the level of detail of the risk management plan should be commensurate with the level of risk associated with the medical device.*
3. *The requirements in ISO 14971:2019 are the minimum requirements for a risk management plan. Manufacturers can include other items such as time-schedule, risk analysis tools, or a rationale for the choice of specific risk acceptability criteria.*
4. *Risk management plan can also applies to the product realization process (design, development and production of the medical device).*
5. *Other elements can apply to the production and post-production phase (such as installation, use, maintenance, decommissioning and disposal of the medical device).*

1.0 Scope of Activities *ISO 14971, 4.4, a)*	This risk management plan applies to the risk management activities, the responsibilities and authorities of those involved, the criteria for risk acceptability, the production and post-production information to be collected and reviewed for

	Advance 101, Battery Powered Digital Blood Pressure Monitor, UA1101, and all risk management activities that are carried out during the entire product life cycle.
	This risk management plan will be reviewed and updated throughout the product life cycle as new information becomes available.

2.0 Device Description *ISO 14971, 4.4, a)*	Advance 101, UA1101, is a battery powered digital blood pressure monitor, for the measurement of blood pressure via upper arm constriction using a pressurized cuff. The device is intended for use in a home healthcare environment
2.1 Product Names	The device is sold under the following trade names: - Advance 101, Battery Powered Digital Blood Pressure Monitor, UA1101 (US & European Market) - Advantus 101, Battery Powered Digital Blood Pressure Monitor, UAL1101 (Latin America)
2.2 Intended Use	The device is designed and manufactured to measure blood pressure and pulse rate of people for diagnosis. It is intended for use on adults only. The device is suitable for home healthcare and is to be used in It is recommended that blood pressure monitoring is conducted while liaison with a qualified physician.

The functions identified below are responsible for the review and approval of this Risk Management Plan.

Notes on Responsibilities and Authorities

1. *Reviewers and approvers of the Risk management plan must be competent and knowledgeable. Training to Risk management procedures is fundamental.*
2. *ISO 14971 does not specify the functions required. This is the responsibility of the manufacturer and should be based on the nature of the device and the risk management procedures.*

3.0 Responsibilities and Authorities ISO 14971, 4.4 a)	Function	
	Function	Technical Expertise
	Risk Management SME	*Knowledge of the risk management process and ISO 14971: application of risk management for medical devices and appropriate regulations*
	Device R&D	*Provides technical knowledge on the operating characteristics and performance of the device*
	Engineering	*Supports the risks management process with process and manufacturing knowledge and experience*
	Operations	*Provides Knowledge of the manufacturing process*
	Quality	*Responsible for the consistent application of procedures*
	Clinical Affairs	*Provide expertise and clinical evaluation*
	Regulatory Affairs	*Reviews for compliance to regulations requirements*
	Medical Expert	*Provides medical expertise, supports literature reviews and other activities*
	Nonclinical	*Directs and executes the investigation and reporting of nonclinical testing*

Notes on Risk Acceptability

1. *For each risk management plan the manufacturer needs to establish risk acceptability criteria that are appropriate for the particular medical device*
2. *It is important to establish the criteria for risk acceptability before starting the risk assessment. Otherwise, the results of the risk assessment could influence the decision when establishing the criteria.*

4.0 Risk Acceptability ISO 14971	The criteria for risk acceptability is established in the policy for determining acceptable risk. The methodology used to evaluate the overall residual risk, and criteria for acceptability of the overall residual risk based on the policy for determining acceptable risk below.

4.4 d)	**Risk Management Policy**
	The Risk Management policy for diagnostic devices is intended to provide safe, reliable and effective products to our customers when the products are used in accordance with specified operating instructions.
	Acceptability of risks is defined in the Risk Management Plan. Risks are identified in the risk management documents. All identified safety related risks are mitigated to as low as possible, where residual risks remain, a risk-benefit analysis shall be performed.
	Where the probability of occurrence of harm cannot be estimated, the criteria for risk acceptability shall be based on the severity of harm alone.
	The evaluation of the overall residual risk is determined upon the review of data and literature for the medical device and similar medical devices on the market which is reviewed by the cross-functional team including medical and clinical expertise.
	Probability:
		Term	Value	Probability per opportunity	Parts per million opportunities
---	---	---	---
Frequent	5	>1/100	>10,000
Probable	4	1/1,000 - 1/1,00	1000-10,000
Occasional	3	1/10,000 - 1/1,000	100-1000
Remote	2	1/100,000 - 1/10,000	10-100
Rarely	1	<1/100,000	<10

Severity:

Term (Severity)	Severity Value (S)	Description
Catastrophic	5	- Patient Death - Destruction of Facility
Critical	4	- Permanent Impairment or life threatening injury – blindness - Destruction of a piece of capital equipment
Serious	3	- Injury or impairment requiring professional medical intervention - Failure of equipment requiring postponement to second surgery - Damage to equipment or facility requiring repair by technicians or contractors
Minor	2	- Temporary injury or impairment not requiring professional medical intervention - Damage to equipment or facility requiring repair by users
Negligible	1	- Inconvenience or temporary discomfort - Delay of start of surgery, or interruption of surgery of less than 30 minutes
None	0	- No harm to patient or user - Delay of start of surgery, or interruption of surgery of less than 30 minutes - Equipment may not work, but no harm to other equipment or facility

Risk Matrix:

	S0 None	S1 Negligible	S2 Minor	S3 Serious	S4 Critical	S5 Catastrophic
P5 Frequent	A	A	A	U	U	U
P4 Probable	A	A	A	U	U	U
P3 Occasional	A	A	A	A	U	U
P2 Remote	A	A	A	A	U	U
P1 Rarely	A	A	A	A	A	U

ISO/TR 24971

- Risks identified shall be assessed for acceptability based on the application of risk estimation and risk analysis
- All residual risks must meet the acceptable residual risk determination criteria. The overall residual risk will be addressed in the risk management report.
- If device data is not available on the probability of occurrence of harm, the acceptance of a risk shall be on the basis of the nature and severity of the harm.
- Determination of acceptable risk for the device is based on applicable standards, comparison of risk from medical devices already on the market and evaluation of clinical data.

Notes on Verification of Implementation

1. *The risk management plan is required to specifies how the two verification activities required completed.*
2. *Verification of implementation of risk control measures can be part of design review, approval of specifications, design and development verification in a quality management system, or other verification activities in a quality management system.*
3. *Verification of the effectiveness of risk control measures can be part of design and development verification in a quality management system. It can require the collection of clinical data, usability studies, etc., as part of design and development validation in a quality management system.*
4. *FMEAs should be developed, reviewed and approved for each product based on company procedures.*

5.0 Verification of Implementation	The verification of implementation of risk control measures are documented in the Failure Modes Effects and Analysis. These include: - **Design Failure modes & Effects Analysis (DFMEA)** - **Process Failure Modes & Effects Analysis (PFMEA)** - **Use Related Failure Modes & Effects Analysis (UFMEA)** - **Design Risk Analysis** - **Risk Identification:** Potential risks are recorded in each FMEA, based on review of the data available, information from similar devices and information from design verification and design validation activities, - **Risk Estimation:** Where applicable, a Risk Priority Number (RPN) shall be used to quantitatively provide risk estimation for potential hazards. The RPN shall be calculated from the Severity, Occurrence and Detection scoring per FMEAs. All identified safety risks must be mitigated to as low as possible. If, where, residual risk remains, a benefit-risk analysis shall be completed. - **Risk Control Measures:** FMEAs shall be reviewed and updated throughout design and development, post launch and over the life cycle of the product Risk control and mitigation actions shall maintained throughout the lifecycle.

	• **Risk Acceptance:** All risk identified shall be reduced to as low as possible with available control measures and considered acceptable. Safety risks that cannot be mitigated to as low as possible must be evaluated against a risk/benefit analysis and must meet the acceptable residual risk determination criteria.
ISO 14971, 4.4, f)	

Notes of Effectiveness Review

1. *It is a requirement for the effectiveness of the risk control measures to be verified.*
2. *The results of this verification shall be recorded in the risk management file such as the Risk Management Report*

6.0 Effectiveness review *ISO 14971, 7.2*	The effectiveness of the risk control measures shall be verified and documented. Verification of the Risk control measures shall be recorded in the risk management report and is subject to approval by a cross functional group. Verification of effectiveness can also be performed over the course of the life cycle or the medical device to ensure the effectiveness of risk control measures remain current and meet the requirements of risk acceptability. The risk mitigation measures in the risk documentation (e.g. PFMEA, UFMEA, DFMEA) shall be reviewed to determine effectiveness. The result of risk mitigation activities shall determine if risks have been reduced as far as possible or if the risk can be reduced further

7.0 Data Collection *ISO 14971, 4.4 g)*	Production and Post production data should take the following sources into account: • Design Changes and Change Controls • Product Quality Review • Clinical Evaluation Reports • Cross Functional Risk Assessments • Complaint Analysis and trending

	• Customer Feedback
	• Management Review
	• Quality Control Data
	• Regulatory Feedback and reporting

Production and Post Production Information *ISO 14971, 4.4 g)*	Production and Post production data shall be made available to the risk management process and cross functional teams responsible for reviewing risk and risk acceptability.
	The processes of the Quality management system shall be utilized to provide data and information that is reliable and current.
	The frequency of review of the collected data and information shall be commensurate with the level of residual risk and severity of risks based on expert review and clinical

ISO 14971:2019 requires that changes to the risk management plan be recorded in the risk management file.

Risk Management References ISO 14971, 4.5	References to the below risk documents shall be maintained in the risk file and shall be listed in the **Risk Management Report**
	• Risk Management Plan
	• Design Risk Analysis
	• Use Related Failure Modes & Effects Analysis (DFMEA)
	• Process Failure Modes & Effects Analysis (PFMEA)
	• Risk Management Report

Revision History	Description
A	*Initial version*

Function	Approver Name	Date
Risk Management Representative	*Approvers signature*	*Date of approval*
Device R&D	*Approvers signature*	*Date of approval*
Engineering	*Approvers signature*	*Date of approval*
Operations	*Approvers signature*	*Date of approval*
Quality	*Approvers signature*	*Date of approval*
Clinical Affairs	*Approvers signature*	*Date of approval*
Regulatory Affairs	*Approvers signature*	*Date of approval*
Medical Expert	*Approvers signature*	*Date of approval*
Nonclinical	*Approvers signature*	*Date of approval*

APPENDIX – Lifecycle Overview

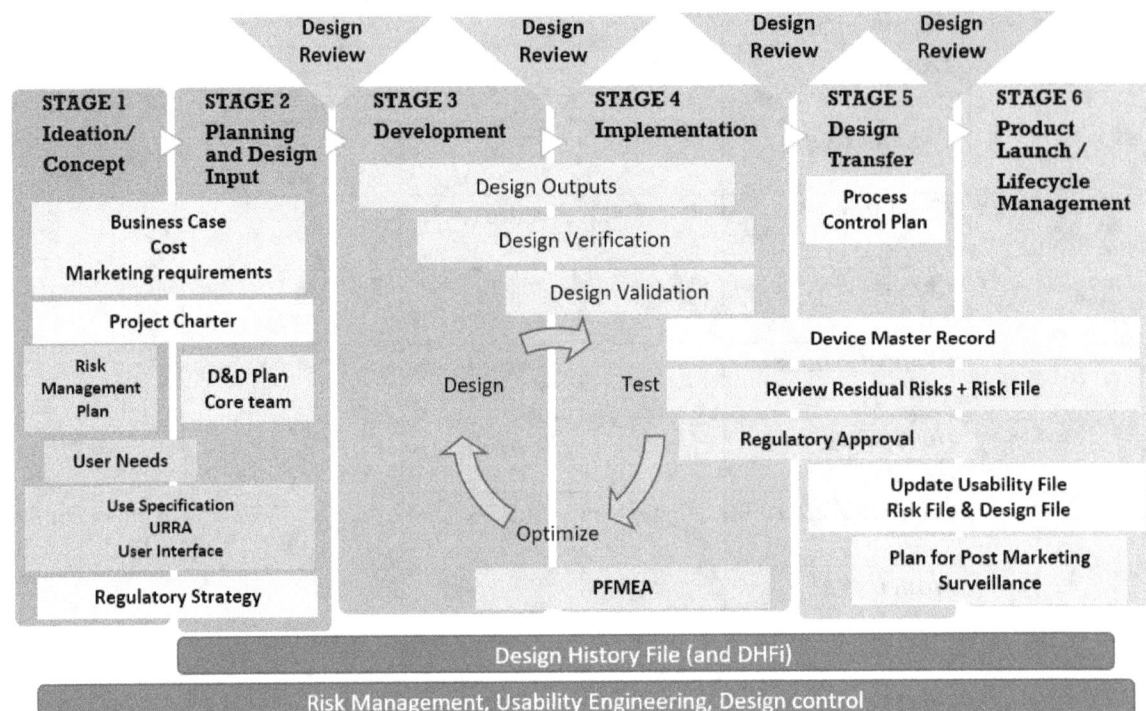

www.ingramcontent.com/pod-product-compliance
Lightning Source LLC
Chambersburg PA
CBHW082106220526
45472CB00009B/2064